Astrotheology
for
Life

Astrotheology for Life:

*Unlocking the Esoteric Wisdom
of Ancient Myth*

David Warner Mathisen

Copyright 2017

David Warner Mathisen

All rights reserved.

No part of this book may be reproduced or transmitted in any form or by any means, electronic or mechanical, including photocopying, recording, or by any information storage-and-retrieval system, without permission in writing from the publisher.

Published by Beowulf Books, Paso Robles, California

Mathisen, David Warner.

Astrotheology for Life: Unlocking the Esoteric Wisdom of Ancient Myth / David Warner Mathisen. – 1st ed.

1. Mythology. 2. Astronomy. 3. Spirituality.

ISBN 978-0-9960590-4-6

Dedicated to

the Divine Spirit in each and every Man and Woman

and to

You

who are reading these words.

May the message inside be a blessing in some way.

CONTENTS

Introduction	1
A Brief Survey of Astrotheology in Modern Times	5
The Genius of H. A. Rey	19
Celestial Cycles and the Ancient System	45
Clothing the Heavenly Cycles in Myth	66
Casting Down and Raising Up: Orion & the Djed	81
The Failed Baptism	105
Cast into the Sea: Crossing the Great Flood	125
Bringing a smile to the lips of the Goddess	142
The Retrieval from the Dead	159
The Divine Twin	191
The Higher Self	207
The Universe Inside	223
Humanity's Ancient History	244
Humanity's Future History	263
End notes	297
Illustration credits	303
Bibliography	306
Index	313

Introduction

The ancient mythology given to the human race holds wisdom vital to our present lives.

Many of us instinctively sense the truth and beauty contained in the world's ancient texts and sacred stories, but when we try to approach the myths themselves we often find them bizarre, circuitous, and difficult to fathom.

Unless we come from an upbringing within a traditional culture unsullied by extensive contact with the implacable spread of "modern western civilization," we find ourselves cut off from the sacred stories which were given to our distant ancestors for their benefit and guidance. They are speaking in a language we no longer understand.

We are like the descendants of immigrants to a new country who were not taught the language of the old country. We may have a longing to reconnect with our heritage, but when we try, we find everything unfamiliar and incomprehensible, and the aching void inside us remains unfilled.

To be sure, there are plenty of voices standing ready to translate the ancient texts -- particularly those speaking the language of literal interpretation of the scriptures -- but as we will soon see, there is abundant evidence that the ancient myths of humanity are speaking a *celestial* language, and unless we are able to listen to them in the language that they are actually speaking, we are likely to misinterpret their message, or miss it altogether. And there are very few today, among all those offering to translate the ancient texts for us, who actually speak that celestial language.

Even worse, if there are translators who do indeed understand the language of the myths, unless we know that language as well, we are at their mercy regarding their translations. In other words, unless we speak the language as well, it is possible that they could mistranslate or misinterpret the ancient myths -- either by mistake or on purpose, for some reason.

There is some evidence that misinterpretation -- even deliberate misinterpretation -- has in fact taken place over the centuries.

Because of this possibility, I would argue that what we need most is not a translator, but rather the ability to go to the ancient stories themselves and listen to them in their celestial language. If we can learn how to speak their language, then this is exactly what we will be able to do. Instead of depending upon an interpreter who knows the language while we do not, we will become capable of speaking the language of celestial metaphor and dialoguing with the myths directly.

That is what this book is designed to help you to do.

Fortunately, as I have attempted to document in my previous books, there is also abundant evidence pointing to the conclusion that all the various ancient myths and sacred stories across the globe and across the millennia, from virtually every culture on every inhabited continent and every inhabited island, use the same worldwide system of celestial metaphor.

They all speak the same esoteric language, clearly recognizable despite significant regional, cultural, and epochal variations.

It is a language that is built upon the motions and cycles of the sun, moon, and visible planets -- and perhaps most of all, upon the cycles of the stars and their constellations. The ancient myths, scriptures and sacred stories of humanity -- including the stories collected into what have come to be called the Old and New Testaments of the Bible -- can be shown to be built upon this same worldwide system.

I have detailed hundreds of examples found in myths from around the world in my most-recent books, the multi-volume series *Star Myths of the World, and how to interpret them*, presently at three volumes reaching to a collective total of nearly 2,000 pages and including hundreds of diagrams and star charts, supporting the conclusion that the same system of celestial metaphor informs them all.

This book, while covering some aspects of the system which have been included in previous books, will add much that is brand-new. It has been conceived with the purpose of teaching you the ancient system in the clearest and most straightforward way possible, with the benefit of everything that I have learned about that system up to this point.

It is also intended to explore more deeply the meaning and purpose of the ancient esoteric message, and its direct application for our lives to this day. It may very well be that the ancient myths and traditions direct us to practices we should be incorporating in our daily lives, practices whose importance we might completely fail to appreciate without the understanding conveyed by the secret language of the ancient wisdom.

And, if you already practice one of these ancient disciplines, you may well find that the wisdom included in the ancient myths can shed new light on what you are doing -- once you understand the language they are speaking and how to converse with the myths for yourself.

Instead of grouping our examination of the ancient myths by separate cultures or traditions, as in the *Star Myths of the World* series, in this volume we will explore certain specific themes or motifs which turn up again and again in myths spanning the globe, and use our investigation of those themes to gain insight into the urgent and beneficial message that the ancient wisdom encoded in the sacred stories may be trying to convey to us.

Pioneering Swedish folklorist Carl Wilhelm von Sydow (1878 - 1952) coined the term "oicotype" (sometimes spelled "oikotype") in 1927 to describe variations upon a story-pattern found in different cultures or different parts of the world. In doing so, he was borrowing a term from botany, describing local or regional variants in a plant species.[1]

Some of the recurring Star-Myth themes or oicotypes we will explore include the motif of the "failed baptism," the "retrieval from the dead," the "forcing of the smile," the "passage through the flood to achieve immortality," the theme of "doubting and withdrawing," and the theme of the "re-establishment of the Djed" (the *Djed* or *Djed-column*, sometimes called the *Tat* or *Tat-pillar* in earlier centuries, is a specific Egyptian manifestation of a pattern that can be shown to repeat throughout many other sacred traditions literally around the globe and across the millennia). By observing numerous such myth-patterns, you will begin to develop the ability to detect the patterns in the world's ancient myths, their connections to certain regions of the sky, and their possible interpretation for yourself.

It is my hope that this book, and the interaction with the ancient myths that it enables, will be a blessing to you.

A Brief Survey of Astrotheology in Modern Times

The study of the evidence that our world's ancient myths, including the accounts in the Old and New Testaments of the Bible, rest upon a foundation of celestial metaphor is frequently termed *astrotheology*.

The term itself will not be found in many dictionaries, although it has certainly begun to gain popular currency, especially following the release of a widely-watched internet video called *Zeitgeist* in 2007, which advances some arguments that the gospel accounts of Jesus are based on stories of earlier solar deities from earlier myth-systems, such as those involving the falcon-god Horus of ancient Egypt.

Unfortunately, pointing out a connection of mythological characters or events to the motions and cycles of the sun (and perhaps at times the moon) is about as far as most popular "astrotheology" analysis goes.

This shallowness of analysis is unfortunate for a few reasons.

First, although it *does* go far enough to cast some doubt upon the validity of the literalistic interpretation of the same myths and stories, noting a few connections between a Biblical text (for example) and the behavior of the sun around the point of winter solstice does not decisively establish the allegorical nature of the stories. One could still argue that the stories contain literal accounts of actual historical and terrestrial figures, whose cultural celebrations down through history were celebrated upon and identified with certain parts of the annual cycle (such as a solstice or an equinox), without having to acknowledge that the persons and events in the stories themselves can be shown to be based on heavenly figures from the outset.

Because the literal interpretation of ancient myth and scripture tends to externalize and thus invert their message, this is a serious concern. We will see that many ancient thinkers, even those who have been presented to us as early Christian leaders, believed very strongly that literalizing the sacred texts was a grave error, emptying them of their true meaning.

Secondly, shallow astrotheology that goes only as far as making connections to the cycles of the sun and the moon does not really provide enough structure to glimpse the purpose of the great esoteric system to most of those exposed to this type of popular astrotheological analysis.

In other words, it goes far enough to undermine the foundations of the literal-historical interpretation (although not far enough to decisively refute the literal and historical approach), while not going far enough to reveal the incredible outlines of the esoteric system and its glorious purpose.

Shallow astrotheology thus serves a primarily *negative* purpose (casting doubt upon the framework of literalism) without a corresponding positive purpose to replace the decrepit structure.

This brings us to a third and related point -- shallow astrotheology as it is usually presented in popular works does not provide much of direct value to our daily lives. Perceiving that there might be a connection between the events in this or that mythological story and the sun's behavior at a solstice or an equinox does not generally translate into specific, applicable knowledge which we can use in our day-to-day journey through this incarnate life.

Once again, this means that shallow astrotheological analysis plays a basically negative role here as well, taking away the perceived applicability of the literalistic interpretation (flawed as that might have been) without offering much insight into the ways in which the ancient myths actually use the awe-inducing heavenly motions and cycles to convey teaching which is simultaneously both practical and profound.

However, if we begin to dig deeper into the great theme of *the celestial metaphor which forms the basis for the world's myths*, we will find an abundance of incredible thinkers and teachers from previous centuries who can help remedy all of the problems described above.

Primary among these is probably Alvin Boyd Kuhn (1880 - 1963), whose penetrating analysis illuminates the outlines of the great cycles of the sun and moon to such a degree that he is in fact able to trace out the mighty esoteric structure and purpose of the world's ancient myths without ever really getting into the specific constellational connections which support and enhance the message that Kuhn demonstrates through his analysis of the cycles of the sun and moon alone.

Of equal importance is the celestial analysis presented in the lectures or sermons of Robert Taylor of England (1784 - 1844). He focused primarily upon the evidence that all the characters and stories of both the Old and New Testaments of the Bible can be shown to be built on celestial metaphor (and in many cases to relate to other ancient myths as well). He is noteworthy for going far beyond the connections to the solar and lunar cycles which often form the bulk of such arguments, and instead uncovering connections to specific constellations in the characters and events described in the Bible episodes.

As we will see, the ability to get to the constellational level of detail in such analysis is absolutely essential.

Robert Taylor actually uses the term "astronomico-theological" in the titles of some of his lectures or sermons, in which he would present to packed audiences in the Rotunda in London his analysis of the celestial underpinnings of familiar Biblical passages and personages.

For example, a sermon delivered on November 27, 1830 at the Rotunda was subtitled "An Astronomico-Theological Discourse on the Temptation of Christ."[2]

In one lecture, he even refers to himself as the master of "this course of astronomico-theological science."[3]

These sermons were published in 1857, thirteen years after Robert Taylor's death -- and his publisher apparently decided that the phrase "astronomico-theological" could be shortened to "astro-theological" instead, and used that term in the subtitle and the introduction to one of the two volumes of Taylor's discourses.

The two collections of his sermons were published in 1857, one under the name *Astronomico-Theological Lectures*, and the other under the name *Devil's Pulpit*. These two volumes will be referenced under those two titles in this volume.

However, the text which this volume will refer to as *Devil's Pulpit* is actually titled in full as follows: *The Devil's Pulpit: or Astro-Theological Sermons, by the Rev. Robert Taylor, B. A.,*

Author of the "Diegesis," "Syntagma," &c., with a Sketch of his Life, and an Astronomical Introduction.

Thus, we see that Taylor himself used the term "Astronomico-Theological," and that after his death his publisher used the shorter term "Astro-Theological." I personally would prefer the term astro*myth*ology, as being more accurate and less prone to misinterpretation, but in deference to Taylor (and to popular usage) will employ the term *astrotheology* in this volume.

In fact, the introduction to the volume entitled *Astronomico-Theological Lectures* tells us that:

> Taylor never approved the title "Devil's Pulpit," which the London publisher has, nevertheless, affixed to all his lectures. He was for naming them, (as the American publisher has their second series) "Astronomico-Theological Lectures."[+]

The genius of Robert Taylor's analysis lies in his facility for connecting specific characters and episodes from the Old and New Testaments of the Bible to specific constellations and celestial cycles. There are many specific identifications where I differ from Taylor's suggestions, but on the whole I have not seen any other proponents of astrotheology who come anywhere close to Taylor's ability to connect to specific constellations. His work thus greatly expands our understanding of the "vocabulary" of the ancient system of celestial metaphor, by introducing hundreds of connections to specific constellations, rather than simply pointing to parallels with the cycles of the sun and the moon.

I am convinced that we must get to the level of *constellational* detail in order to develop our own "vocabulary" to the point at which we can approach the myths for ourselves and ask them what they mean -- and at which we can then hope to understand their answer.

Although the system of correspondence between the myths and the constellations must have been known for thousands of years -- as we will see from some of the surviving artwork from ancient times which clearly displays evidence of understanding of this system -- I am unaware of any writings prior to the work of Robert Taylor who attempted to articulate it to the degree of detail which we find in Taylor's sermons.

The works of Volney and especially of Charles-François Dupuis (both of France) deserve mention in their exploration of the same broad theme, but once again they focus primarily upon the cycles of sun and moon, and when they do address connections between the myths and the stars, the analysis lacks the detailed level of constellational specificity which we find in Taylor's analysis.

Their particular genius was to show the parallels between the episodes contained in the Bible stories and the myths of ancient other cultures, and arguing some connection or even common origin based on the undeniable correspondences. As some of those other cultures (such as ancient Egypt and ancient Sumer and Babylon) can be shown to have been in full flower before the alleged Exodus described in the Pentateuch, the argument can be made that the Biblical texts are not the original source of these stories or patterns.

Volney (1757 - 1820) and Dupuis (1742 - 1809) were both alive at the same time as Taylor, Volney being 27 years of age when Taylor was born, and Dupuis being an additional 14 years older than Volney. Their most influential books were *The Origin of All Religious Worship* (*Origine de tous les Cultes*) by Dupuis (published in France in 1795) and *The Ruins: or, Meditations on the Revolutions of Empires* (*Les Ruines, ou méditations sur les révolutions des empires*) by Volney (published in France in 1791 and later translated into English by Thomas Jefferson).

Both men posit a common origin in the natural phenomena and especially the heavenly phenomena for all religious worship -- and generally take a sort of evolutionary position on its development, with Volney specifically positing "star worship or sabaism" as one stage along the journey (a position that Taylor also appears to advocate, but one which has some strong arguments against it, as we shall explore).

It is certainly possible that the writings of one or both thinkers influenced Taylor's own development of "astronomico-theological" analysis. A short biographical sketch included at the beginning of the *Devil's Pulpit* collection tells us that Taylor had been persuaded to become ordained in the Church of England by a minister who observed Taylor's "strong religious feeling and great powers of oratory," but that after Taylor had graduated from St. John's College in Cambridge and had been preaching for some time, first in London and afterwards in Surrey, it was "a free enquirer, a tradesman in Midhurst" who, "by the loan of books and conversation first awakened his skepticism."[5]

No further detail, unfortunately, is provided regarding the titles of those books loaned to the Reverend Taylor by the anonymous "free enquirer" of Midhurst.

Taylor's detailed and thorough exposition of the parallels between specific Biblical episodes, from Adam and Eve to the birth in the manger and the visit of the Magi, from the twelve tribes of Israel and the vision of Ezekiel to the twelve disciples and the astronomical significance of the four "gospel creatures," provides a devastating critique of Biblical literalism.

His groundbreaking work may have been partly responsible for the publication in subsequent decades of books by Christian literalists who tried to argue that the presence of a "gospel in the stars" supports and reinforces the evidence for the historicity of the events described in the Biblical scriptures.

In 1862, a work entitled *Mazzaroth: or, the Constellations* was published by Frances Rolleston (1781 - 1864), which takes a literalist view of the Bible but outlines the connections to the celestial figures of the zodiac and other constellations. And in 1884, Joseph A. Seiss (1823 - 1904) published *The Gospel in the Stars: or, Primeval Astronomy*, which also alleges that the connections to the constellations provide additional evidence supporting the interpretation of the Biblical stories as literal terrestrial history.

Rolleston argues that the accounts of later Biblical events, and especially of the life and ministry of Christ, were placed in the heavens during "the first age of mankind," probably during the life of Seth the son of Adam and Eve, in order to "express the promises and prophecies revealed to Adam, Seth, and Enoch."[6]

Seiss, consciously following the work of Rolleston and building upon her arguments (as well as mentioning Albertus Magnus, a Bavarian bishop and later saint, who lived from about 1200 to 1280, and to whom a work known as *Speculum Astronomiae* is attributed, although scholars now believe it may not be correctly attributed to him after all), declares that in the stars God has given "a revelation of truths and hopes precisely as written in our Scriptures, and so fondly cherished by all Christian believers."[7]

He also cites the author of *Speculum Astronomiae* who said, back in the 13th century: "the mysteries of the Incarnation, from the Conception on to the Ascension into heaven, are shown us on the face of the sky, and are signified by the stars." [8]

As a literalist Christian minister, Seiss of course interprets this sentence as meaning that the stars show us a record of a literal and historical Incarnation of the divine Christ child who is named Jesus in the gospel stories -- but note that the sentence quoted does not necessarily indicate that this is what it means. The sentence as translated above and as quoted by Seiss could very well mean that the Incarnation in a general sense – the incarnation, that is, of each and every man and woman who takes on a body, and thus "incarnates" -- and the same for the Ascension into heaven (all of these stories being interpreted esoterically) -- are signified *symbolically* in the heavens.

Literalists have generally abandoned the line of argument exemplified by the work of Rolleston and Seiss in the above texts, because, although Rolleston and Seiss argue that the correspondence between the Biblical stories and the ancient heavenly constellations supports a literal and historical interpretation of those stories, the same evidence could very well be seen as arguing that the stories themselves were never recounting literal, terrestrial history at all, but rather stories based on the stars (the same as all other ancient myths which appear to follow the same patterns). By the twentieth century, this would certainly be the way such evidence would be likely to be interpreted, rather than as supporting the literal historicity of the events and characters which can be seen to be reflected in the celestial figures.

One of the most significant and influential works to take up this subject in the twentieth century is undoubtedly the 1969 text written by Giorgio de Santillana and Hertha von Dechend, entitled *Hamlet's Mill: An Essay on Myth and the Frame of Time*. In it, the authors (both university professors) argue that all the world's ancient myths appear to share some ancient, now-forgotten pattern, which is somehow related to the motions of the sun, moon, stars, and visible planets, and to the mighty celestial cycles, particularly the slow inexorable cycle known as the precession of the equinoxes.

Hamlet's Mill does provide some proposed one-for-one linkages between individual mythical characters or episodes and specific constellations, and the authors also include extensive citations and footnotes to previous scholars and philosophers who suggested such linkages in earlier decades or centuries.

However, while they do an admirable job of establishing the fact that the ancient myths and scriptures are in fact speaking a celestial "language," the actual range of their "vocabulary" leaves much to be desired for specific "astrotheological" correspondences. Aside from a few connections to constellations, the authors of *Hamlet's Mill* usually fall back upon possible connections to certain visible planets (especially the planet Saturn), and to connection with the phenomenon of precession (admittedly a vitally important celestial cycle and one that is undeniably active in many ancient myths).

Rather than provide an extensive "glossary" of explicit correspondences between one mythical character and his or her celestial analog, the authors of *Hamlet's Mill* more often give coy and ambiguous hints, often saying in effect, "this is important, and clearly celestial, but it is too soon to reveal the actual celestial correspondence," and implying that they will come back to it later. Unfortunately for the reader, they rarely do revisit it, and if they do they are usually just as indecisive (see the several places in the text in which they say, "We must leave it at that" or, "one may guess who [. . .]," or "Since this is not a manual on the Epic of Gilgamesh, this whole affair of the plant, the diving, the fateful bath in the well, must stand as it is, [. . .]" and "The scholar is already in a hopeless tangle, and Lord help him" and many other similar deflections). [9]

However, two of the correspondences which the authors of *Hamlet's Mill* do spell out explicitly were decisive in my own journey of realization that the scriptures of the Bible were not originally intended to be understood as literal, terrestrial history -- and started me on my own quest to try to understand this celestial "language" in a more systematic and comprehensive manner.

I have written about these specific Biblical examples from *Hamlet's Mill* in some of my previous books, such as *The Undying Stars* (2014) and *Star Myths of the Bible* (2016). These examples are the Samson story and specifically the use of a jawbone as a weapon (which *Hamlet's Mill* connects to the outline of the Hyades in the constellation Taurus) and the vision described in Revelation chapter 9 (in which the scorpion-tailed horsemen with long hair and crowns, beside a column of smoke arising from the "bottomless pit" are argued to be an allegorization of the constellations Sagittarius and Scorpio on either side of the brightest part of the Milky Way column, which is described as the smoke arising from the pit --

and, as I realized later, we could add Corona Australis to that group as well).[10]

However, we must not be too hard on the authors of *Hamlet's Mill*: they freely admit that they are standing before the vast ruin of an ancient system that has been lost "so utterly that in no Western language is there a word to express it."[11] The genius of their text is to show beyond doubt that this ruin indeed exists, even if it is almost entirely buried beneath the sands of time, and also to show that it is worldwide -- and the genius of the text lies most especially in the sense of awe which their survey invokes, as they move from one perspective to another within this vast ancient ruin.

The authors of *Hamlet's Mill* very clearly demonstrate that this vast worldwide construct must be the product of a deep and extremely sophisticated knowledge, one whose thoughts run in channels that are completely alien to our own (speaking from the perspective of "the modern consciousness" as they saw it from their perspective at the end of the 1960s).[12]

Von Dechend and de Santillana conclude "that myth is neither irresponsible fantasy, nor the object of weighty psychology, or any such thing. It is 'wholly other,' and requires to be looked at with open eyes. This is what we have tried to do."[13] At this, I believe, they have succeeded quite admirably -- not to explain how it all fits together, but to direct our wondering gaze at this awesome, interconnected, worldwide tangle of myth . . . and to force us to admit that it follows a logic and a speaks a language that is "wholly other" to our usual way of thinking and defies the categorization with which it is usually described by conventional academia.

The passage from an essay published by de Santillana in 1959, which the authors of *Hamlet's Mill* include in their introduction, is hard to rival for its evocation of the sense of mystery and feeling of wonder these ancient ruins can inspire:

> The dust of centuries had settled upon the remains of this great world-wide archaic construction when the Greeks came upon the scene. Yet something of it survived in traditional rites, in myths and fairy tales no longer understood. Taken verbally, it matured the bloody cults intended to procure fertility, based on the belief in a dark universal force of an ambivalent nature, which seems now to monopolize our interest. Yet its original themes could flash out again, preserved almost intact, in the later thought of the Pythagoreans and of Plato.
>
> But they are tantalizing fragments of a lost whole. They make one think of those "mist landscapes" of which Chinese painters are masters, which show here a rock, here a gable, there the tip of a tree, and leave the rest to imagination. Even when the code shall have yielded, when the techniques shall be known, we cannot expect to gauge the thought of those remote ancestors of ours, wrapped as it is in its symbols.
>
> Their words are no more heard again
> Through lapse of many ages . . . [14]

Their text has been a seminal one, spurring numerous researchers to begin examining this "great world-wide archaic construction" more closely, and to offer possible frameworks within which the various remaining structures -- ruined arches and tumbled-down observation towers and shattered walls -- might fit.

And, as mentioned above, the clear demonstration in *Hamlet's Mill* that the stories of the Bible can be shown to be part of this same mysterious world-wide system was seminal to my own realization that the scriptures which I had been taking as literal and historical for nearly twenty years of my adult life were speaking a very different language than the one I had been told they were speaking.

This revelation caused a major paradigm shift impacting many areas of my life and most of the patterns into which I had previously been attempting to fit the data of the outside world, and spurred an intense need to try to understand more of the shape and purpose this mysterious world-wide ruin which de Santillana and von Dechend (and other explorers before them) had reported finding forgotten and neglected amongst the jungle overgrowth and the blowing desert sands.

In that quest, I had assistance from a source which could not have been known to Alvin Boyd Kuhn, or to Robert Taylor, and one which appears to have been also unknown to Giorgio de Santillana and Hertha von Dechend even though by the 1950s and 1960s it would have been available to them: the inspired constellational-outlining system of an author more famous for the children's books which he and his wife Margret wrote and illustrated, centered on the exploits of an irrepressible little primate named George (who is usually called a *monkey*, but since he doesn't have a tail he should probably be classified as an *ape*).

To this system of outlining the figures in the stars -- a system which appears to track very closely with whatever system was inherited by the most ancient civilizations whose art we can examine today, and a system which provides an essential key to unlocking many of the celestial correspondences which cannot be perceived without it – we shall now turn.

The Genius of H. A. Rey

Hans Augusto Rey (1898 - 1977) and his wife Margret Rey (1906 - 1996) were married in 1935, and in 1941 they published the first *Curious George* storybook, in New York. It was an instant success.

I remember reading the Curious George stories as a child, with their memorable illustrations (especially the story in which George decides to turn on the water in the apartment while the Man in the Yellow Hat is away, flooding the building), but even more than that I remember reading H. A. Rey's books on the stars and constellations -- first *Find the Constellations*, and soon after that *The Stars: A New Way to See Them*. I had access to both of these wonderful constellation guides at such a young age that I cannot actually remember when I started reading them.

As I have explained in previous publications, H. A. Rey explains in *The Stars: A New Way to See Them* (first published in 1952, and still in print in slightly updated editions to this day) that he was frustrated with the way the constellations had been represented in books purporting to provide a guide to the night sky. He explained:

> There are, of course, plenty of books about the subject, and they do very well in most respects. But in one important point they seem to fail us: *the way they represent the constellations.*[15]

The trouble, he noted, was that almost all the diagrams of the constellations were unhelpful in one of two different ways.

The first way that constellations have been unhelpfully depicted, most common in previous centuries up to and including the end of the nineteenth and even the beginning of the twentieth, shows the heavenly figures by superimposing flowery artwork over the stars, which may help us to imagine what the constellations are supposed to represent, but which are completely superfluous to what we can see in the night sky and which are virtually useless for trying to locate the figures when we go outside to look for them. These figures do not even show the "lines" which we should imagine between the individual stars of the constellation.

As H. A. Rey says: "Some books show, arbitrarily drawn around the stars, elaborate allegorical figures which we cannot trace in the sky."[16]

The other and almost "opposite" problem characterizes the method used in more recent decades, beginning in the twentieth century, which does indeed give us lines to imagine between the stars in order to envision a figure -- but the connecting lines in virtually every case seem to have been perversely conceived in the most unhelpful way imaginable, almost as if the designers of the system have been attempting to *conceal* the constellation instead of revealing it!

Lamenting this problem -- which apparently spurred him to publish his own system -- H. A. Rey writes:

> Others, most of the modern ones, show the constellations as involved geometrical shapes which don't look like anything and have no relation to the names. Both ways are of little help if we want to find the constellations in the sky -- yet this is precisely what we are after.[17]

He might have added that, not only are they unhelpful for finding the actual constellations in the night sky, but these two unhelpful systems are also of little use for perceiving the connection between the stars and the figures found in ancient myth. We will illustrate both of the problematic systems and also H. A. Rey's brilliant solution in the pages which follow -- along with ancient artwork which demonstrates that the connection between the myths and the stars must have been understood at some extremely ancient point in time, and that this connection has been incorporated into artwork down through the millennia.

Below we see an example of the first unhelpful type of system for envisioning the constellations, in which "elaborate allegorical figures" have been "arbitrarily drawn around the stars." This diagram is for the constellation Virgo, from a collection entitled *Urania's Mirror* published in 1825 and illustrated by Sidney Hall (1788 - 1831):

We can readily appreciate H. A. Rey's frustration with the lack of utility such an image would have as an aid to actually locating the constellation Virgo in the night sky.

No one can simply look up and see those outlines in the heavens, because they exist only in the imagination of those who already know the traditions surrounding the constellation as a representative of a beautiful woman or maiden. This image will not enable anyone to be able to actually locate the constellation Virgo, even if staring directly at the stars which make up the constellation.

Below is an example of the opposite problem: what H. A. Rey calls "shapes which don't look like anything." This is an image of the stars of the constellation Virgo, with connecting lines added according to one modern methodology. Remembering such an outline, and finding it in the sky, would be a real challenge -- especially as the outline bears absolutely no resemblance to the Maiden for whom the constellation is named:

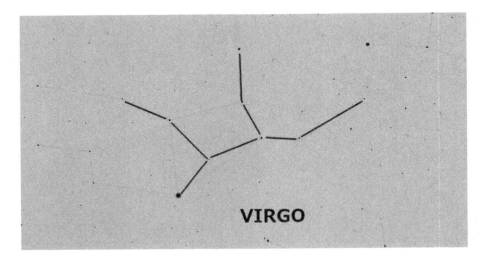

And below is the exact same section of the sky, showing the exact same stars, but this time connected as envisioned by H. A. Rey:

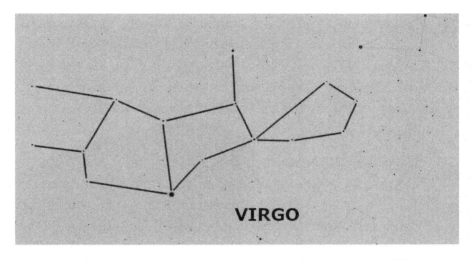

We can now make out the outline of a recumbent woman. This at least gives the suggestion of the constellation's namesake, the Virgin. We can remember this form and find it in the sky, in a way that would be virtually impossible to do with the conventional version, which resembles nothing more than a random collection of sticks.

Note that there *are* other suggested ways of connecting the stars of Virgo which you might find, but most of these are just as useless as the one selected and shown above -- in other words, there may be two or three contenders for the "conventional" outline of any given constellation, but unless they are Rey's version, they seem to be contending for the prize of "most unlike the shape you are supposed to be looking for."

Not only is H. A. Rey's version much more memorable and much more useful for actually locating the constellation Virgo in the sky, but his outline also preserves features or characteristics of Virgo which are important for actually matching her to myths and sacred stories from around the globe.

For example, Virgo's outline features a distinctive "outstretched arm," which is marked by the bright star Vindemiatrix and which turns out to be one of the most distinctive aspects of the constellation, playing an important role in many Star Myths in which a character based upon Virgo appears.

H. A. Rey's Virgo also has an elongated head -- and interestingly enough there are myths from the indigenous Aborigines of Australia which feature a figure known as the "Bandicoot Woman," who can be shown fairly definitively to correspond to Virgo. Bandicoots belong to a genus of marsupial of which numerous species are found in Australia: their somewhat mouse-like head is elongated and comes to a pointed nose, somewhat reminiscent of this suggested outline of the constellation.[18]

Similarly, the Maya have a myth about a goddess named Xmucane, who is associated with the coati, which is a large ring-tailed animal found in the jungles of Central and South America.[19]

The outline of Virgo is reminiscent of a woman seated on a chair or a throne -- and this is another distinctive aspect of the constellation which will appear in myth. Sometimes goddesses who are associated with Virgo will be seated on a throne flanked by one or more lions, or will ride in a chariot or cart which is drawn by one or more lions -- no doubt due to the fact that the constellation Virgo follows closely behind the constellation Leo the Lion.

The shape of the outline of Virgo envisioned by H. A. Rey obviously corresponds closely to that envisioned by ancient cultures and preserved in their artwork: below is an image of the Pythia who according to Greek myth would sit upon a tripod at the sacred Oracle of Delphi. Note the outline of her figure, as well as her distinctively Virgoan *outstretched arm*:

It bears repeating that the clear similarity of this and other pieces of ancient artwork from around the world to the outline of the constellation Virgo would be well-nigh impossible to spot using the conventional outlines offered to us, which "don't look like anything."

25

Before leaving this brief discussion of the constellation Virgo and the merits of the H. A. Rey system of outlining the constellation versus the conventional systems, it should also be pointed out that the form of the constellation can not only be envisioned as seated upon a throne or a tripod, but can also be envisioned as a woman lying on her back with her legs elevated in the air, and in fact with her feet spread apart -- exactly as if about to give birth.

This aspect of the constellation Virgo also features in numerous myths from around the globe. The zodiac constellations which follow immediately behind Virgo are first Libra the Scales (a very faint constellation) and very close behind Libra the dazzling form of Scorpio (frequently called Scorpius to distinguish the constellation from the zodiac sign, especially in previous centuries, but now both the sign and the constellation are often referred to by the same name of Scorpio).

The close proximity of Scorpio to the recumbent form of Virgo with feet elevated and spread apart gives rise to the many myths in which a woman or goddess gives birth to a monstrous serpent or a hydra -- or even a child with multiple heads.

In some of the versions of the sacred stories involving the demi-god Maui, whose mythology is found across the Pacific, the baby Maui has eight heads (a common aspect of the constellation Scorpio in various myths, in which the constellation is envisioned as having multiple heads, sometimes numbering seven, eight, or nine heads). In some versions of the myths, this horrifying aspect of the infant causes one or both of his parents to cast the newborn Maui into the ocean.[20]

Note that the constellation Scorpio is located partly within the shining band of the Milky Way -- which is sometimes envisioned in the myths as a great stream, and even as the Ocean Stream. Thus the correspondence of Scorpio with the stories of throwing Maui into the ocean make even more sense when we understand this aspect of the constellation's situation.

In the book of the Revelation included in the New Testament canon, we are told of a woman about to give birth, and a great red dragon lying in wait to devour her child (Revelation 12). This again is undoubtedly founded upon the relative locations of Virgo (who appears like a "woman in travail" or in the pangs of labor, about to give birth) and Scorpio, lying beneath her upraised legs. The red dragon, in fact, is described as having seven heads in that passage (Revelation 12: 3).

In other myths, the curving shape of Scorpio is actually envisioned as a ship (it almost resembles the shape of a Viking ship) or as a boat -- and I am convinced that the many myths from around the world in which a baby is placed into a reed boat immediately after being born refer to this same part of the night sky, and the fact that Virgo can be envisioned as a woman in labor, about to give birth.

Figures who are placed into reed baskets and set adrift include Moses in the Old Testament, Sargon in the legendary history of Akkad (in particular in an Assyrian account in which Sargon says he was set in a basket of rushes, sealed with bitumen), Karna in the Sanskrit epic Mahabharata of ancient India, and the child Hiruko (the Leech-child) in the sacred Kojiki of ancient Japan (among others -- some versions of the Oedipus story may involve his being cast adrift upon a river as well).

All of these myths undoubtedly relate to the proximity of Virgo to Scorpio and the Milky Way, and to the fact that Virgo's outline makes her appear to be in the act of delivering a child. Also, immediately above Scorpio and also along the edge of the heavenly "river" of the Milky Way we find the constellation Ophiucus, which resembles a rectangular box with a triangular lid – and which may well furnish the outline for the "ark" in which some of these Star Myths describe the newborn infant being cast adrift. Virgo with her upraised legs is directly in front of both Scorpio and Ophiucus.

Again, without the outline furnished by H. A. Rey, we would not be able to see that Virgo can be envisioned as a woman in labor, about to give birth (or perhaps just having given birth). We certainly cannot see that important characteristic of Virgo using either of the two non-Rey alternatives shown above – neither the elaborate and flowery outline of Virgo depicted in *Urania's Mirror*, nor the spare, sparse, insect-like jumble of lines from the conventional modern outline, give us any clue to this aspect of Virgo's mythological analogs.

It should be stressed that, although the constellation Virgo will very frequently appear in myth as a female character or goddess, this will not always be the case. There are times when male figures will also correspond to Virgo -- and indeed, virtually all of the constellations can appear at different times as either male or female (sometimes within the very same myth, in fact).

For more specific examples of this aspect of the ancient system, see my multi-volume series *Star Myths of the World, and how to interpret them* (there are examples in both the Gilgamesh series of myths and in the story of Adam and Eve in which a male character takes on an aspect of Virgo, and there are also many examples in ancient art in which male characters and gods are depicted in postures which are clearly reminiscent of Virgo, such as a famous depiction of the god Dionysus or Dionysos riding on a cheetah or leopard).

For another example of a constellation which frequently appears in ancient myth as a female character and at other times as a male character, we will briefly look at the zodiac constellation of Sagittarius -- another set of stars for which the outline of H. A. Rey is vastly more helpful than the conventional outline system found on Wikipedia or on most smartphone apps.

Below is an example of the "elaborate allegorical figures arbitrarily drawn around the stars," common in previous centuries:

We know that Sagittarius (whose name means "the Archer") is typically envisioned as a centaur wielding a bow, and the artist

29

has provided an outline to reinforce this idea – but much of the figure has no relation to the stars themselves (note that there are no stars near the rear hooves of the horse, for example, or the tail).

Good luck finding the constellation in the night sky based solely upon the imaginative outline above!

Unfortunately, the "modern" outlines (two versions of which are illustrated on the following pages) are not much better for trying to locate Sagittarius in the sky. In fact, the first one presented is actually even less helpful (if possible) than the "elaborate allegorical figure" shown above.

Again, we must wonder if someone is deliberately connecting the stars using the most illogical method possible, and then presenting their abstract patterns for hopeful stargazers. The first one shown does not even preserve the outline of Archer's bow:

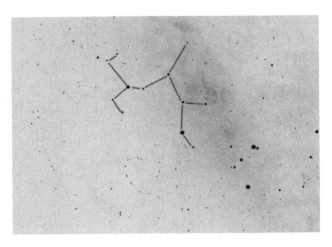

Simply useless: nothing recognizable has been left of the constellation that can possibly identify a celestial Archer. Below is a second modern version, slightly better than the first but still too abstract and random to be of much service (it does, however, preserve the outline of the "teapot" formed by the brightest stars of the constellation):

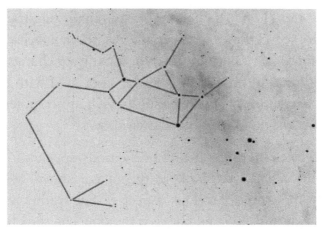

And now, after marveling at the unhelpful versions still in use to this day, have a look at the inspired outline offered by H. A. Rey in *The Stars: A New Way to See Them*:

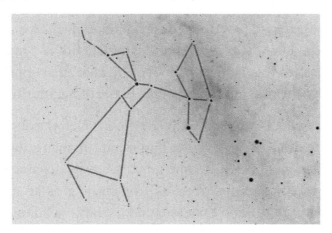

In each of the above star-charts, the exact same set of stars is depicted. The only difference is the choice of which lines to envision between which stars. The superiority of the H. A. Rey method for outlining the constellation is inherently obvious.

With that shape in mind, we can actually go out into the night sky and locate Sagittarius, during the months when the constellation is visible.

Additionally, the H. A. Rey method of outlining Sagittarius enables us to see certain characteristics of the constellation which turn out to be vitally important for identifying characters in the world's ancient myths who may be associated with this particular constellation -- characteristics which are completely obscured using the other methods shown above.

For instance, note that the posture of the body appears to be moving to the left (as we look at the image above) while shooting towards the right. This aspect of Sagittarius, as if walking one way and "looking back" in the opposite direction, will turn out to be extremely important in some of the myth-patterns we will examine in this book, patterns which span cultures, continents, and indeed millennia.

Another characteristic of the constellation, which we can perceive in the H. A. Rey version of the outline but not in any of the others, is the distinctive "plume" or "feather" that appears to wave above the triangular head of the figure of Sagittarius.

Yet another aspect is the long "skirt" or dress which the outline seems to suggest. This will be important in many myths involving this figure. In some myths, male characters associated with Sagittarius dress as women or girls at some point, such as both the god Dionysus and the hero Achilles.

Keeping the above characteristics in mind, look now at this image of the goddess Artemis, slaying the unfortunate hunter Actaeon (or Acteon), from the magnificent "name vase" of the artist known to scholars as "The Pan Painter."

Not only does the artist's depiction of the goddess Artemis correspond very closely to the outline of Sagittarius in the heavens, but the figure of Actaeon corresponds to nearby Scorpio, and in exactly the correct relation to Sagittarius:

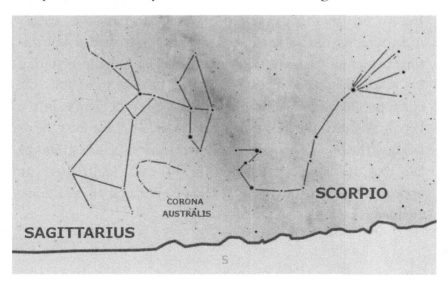

The correspondence between the ancient red-figure image on the bell krater (thought to date to about 470 BC), itself a masterpiece of composition and the human form, and the outline of Sagittarius in the sky is simply undeniable.

The goddess appears to be walking in one direction, while turning around to shoot back in the other direction. She holds her bow at the same height and angle as the outline of the bow in the constellation Sagittarius. She wears an ankle-length dress or skirt, and her feet are positioned in accordance with the outline of suggested by the stars of the constellation. She even has a quiver or sword slung over her shoulder which juts up to the left of her head as we face the image, corresponding to the "feather" or "plume" in the outline of Sagittarius:

It is correspondences of this kind which cause me to wonder whether H. A. Rey was privy to some extremely ancient tradition of envisioning the constellations, or if by some inspiration he happened to come up with the very same outlining system used by the ancient masters.

Below is another example of ancient artwork clearly incorporating the same constellational characteristics, but this time depicting the scene of Odysseus slaying the suitors during his vengeful homecoming. You should be able to identify the correspondence to Sagittarius in the artist's depiction of the hero's posture, the position of the feet, and the way he is holding the bow. This example is provided to demonstrate the fact that the same constellation (in this case Sagittarius) can appear in different myths as a female character or as a male character:

In case the correspondence to Sagittarius in the outline of Odysseus is not obvious in the above illustration, the ancient artist has also included outlines which evoke three adjacent zodiac constellations (labeled above): Scorpio, Libra, and Virgo.

Additionally, the artist has added a flowering design between Odysseus and the three suitors, corresponding to the Milky Way band which rises up between Sagittarius and Scorpio in the sky. All of the figures in the illustration are in the correct positions to correspond to the positions of the constellations as they are seen in the heavens, relative to one another and to the brightest and widest part of the Milky Way galaxy.

Let's look at just one more example which will again demonstrate the absolutely vital importance of H. A. Rey's system of envisioning the outlines of the constellations: the constellation Hercules. Below is an image from *Urania's Mirror*.

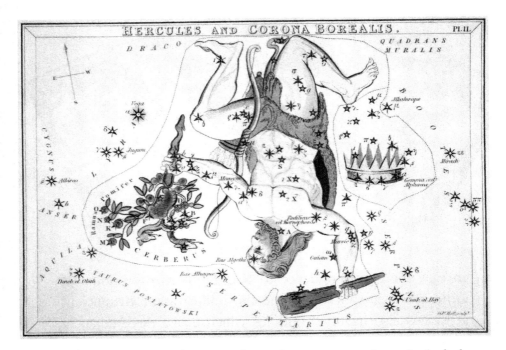

Oddly enough, in this case the artist appears to have included some of the conventions of the outline of Hercules which correspond to the way that Hercules-related figures will often appear in myth and in ancient artwork (from around the globe) – but because the artist is basically drawing-in a figure arbitrarily, with little relationship to the actual stars themselves, the figure of Hercules is actually inverted in relation to the stars which form this outline!

As we will see when we get to the outline as proposed by H. A. Rey, the stars actually do form the image of a powerful figure (usually but not always envisioned as a male figure) in a deep lunge or kneeling posture, with one hand overhead brandishing a massive club and the other hand reaching downward and forward. The artist in this case is probably familiar with this convention -- but the stars which create this outline are actually "right-side up" compared to the way the artist has drawn his ornate figure (the downward-reaching hand is towards the crown shown in the illustration, which corresponds to the position of Corona Borealis).

Below is an example of a "modern" outline connecting the stars of the constellation Hercules. Once again: useless for finding the constellation in the sky, and useless for seeing the distinctive characteristics of Hercules that will help us to identify Hercules-figures in Star Myths around the world, and in ancient artwork.

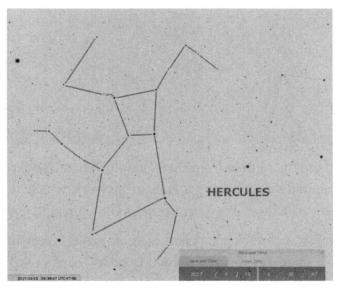

As you can see, none of the distinctive characteristics of Hercules are preserved in the above modern outline -- no deep lunge is visible, we cannot see the hero's upraised arm and menacing club, or his downward-reaching lower arm (which reaches towards the Crown, which is just above the word "Hercules" in the above image).

Below is the outline of the constellation Hercules as suggested by H. A. Rey. The superiority to the completely useless conventional outline is readily apparent. Further, all the traditional elements are clearly identifiable: the deep lunge with bent rear leg, the menacing club or sword raised overhead, and the downward reaching lower arm:

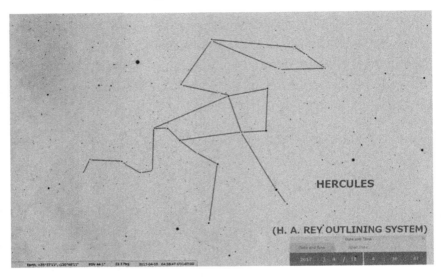

Once again, the outline from the system of H. A. Rey appears to correspond very clearly to ancient artwork which depicts characters from myth who correspond to this constellation.

For example, below is an ancient piece of pottery showing the outline of Hercules during a scene from his conflict with the Amazon women. In this scene, all the aspects of Hercules as the constellation appears in the sky seem to have been incorporated into the ancient pottery art by the original artist. Note the deep bend in the rear leg, as well as the upraised club (or sword) of the hero Hercules, or Heracles; note also the downward-reaching lead arm.

There are numerous surviving pieces of pottery from ancient Greece in which the artist chose to depict Heracles, or Hercules. Most display clear celestial references, hidden though they be to those unaware of the system. The Hercules-related figures on the following pages can be double-checked by turning back to the basic Hercules outline shown on the previous page: deep lunge, downward reaching arm, menacing upper-arm with club or other hammer-like weapon overhead.

The correspondence to the outline of the constellation Hercules in the artwork on this ancient vase is unmistakable. The right arm is raised over the head in exactly the same manner, brandishing a weapon (for figures related to the constellation Hercules, this weapon is often a huge club, a mace, a large sword, or even a thunderbolt). The lower arm is reaching forward towards the head of the female Amazon warrior (note that the curved plume on her helmet corresponds closely to the location of the Northern Crown in the sky, which is also an arc-shape). The rear leg of the hero is extended backwards. The head is square-shaped, and looking slightly downwards (just as the head of the constellation appears slightly bowed-downwards in the sky).

The correspondences between H. A. Rey's outline and the ancient art depicting the hero Hercules (or Heracles) is striking.

The constellation Hercules appears in other myths, playing the role of countless different characters (usually male, but not always). Below is another example, this time from ancient Mesopotamia -- an impression from a cylinder-seal, showing a figure thought to be either Gilgamesh or Enkidu, wrestling a lion:

Again, the correspondences to the constellation's outline are undeniable. The deep bend in the rear leg at the knee, creating the deep lunging position, are emphasized even more strongly in this ancient artwork. The body is also tilted forward in almost the exact same angle as the angle of the constellation in the sky.

In the above artwork, the portion of the constellation which is usually envisioned as a sword is instead envisioned as the lion which Gilgamesh (or Enkidu) is wrestling. In myths in which a Hercules-figure wrestles a lion, the ancient story usually emphasizes that the hero does so *without using his sword* or any other weapon, but instead *using his bare hands*. The reason for this is probably because the stars usually envisioned as the club or the sword in this constellation are now being envisioned as the lion which the hero must wrestle -- without his weapon because the stars that are supposed to be the *weapon* are now the *lion*.

Sometimes, the stars which make up either the extended rear leg of the constellation Hercules, or the stars which arch over the head of Hercules, are envisioned as something completely different -- even as the tail of a monkey. The god Hanuman of ancient India is almost certainly connected to the constellation Hercules. Below is an image of Hanuman, in which the artist has chosen to depict the god in a posture which is clearly evocative of the outline of the constellation Hercules in the sky.

In the ancient Sanskrit epic Mahabharata, the figure most clearly displaying characteristics which suggest the constellation Hercules is the mighty Pandava named Bhima. Not only does Bhima exhibit tremendous strength (a frequent characteristic of figures associated with this constellation) but his chosen weapon is an enormous mace.

Not only that, but Bhima is told by Hanuman himself that the two are related -- which adds further support to the suggestion that Hanuman is also related to the constellation Hercules.

The image below of a sculpted relief depicting Hanuman shows characteristics related to the outline of the constellation Hercules, as envisioned by H. A. Rey:

Here again, the rear leg is extended outwards and deeply bent, which are distinctive characteristics suggesting the constellation Hercules. Also, the right arm is raised and the left arm is angled downwards. In this case, the tail arches over the back and then over the head of Hanuman, just as the great sword arches over the head of Hercules in the sky.

The similarities to the constellation are unmistakable -- but only to the constellation as envisioned by H. A. Rey's inspired outline system.

Somehow, H. A. Rey appears to have either known the ancient system, or to have come upon it through his own inspired genius. Perhaps the system was preserved in secret down through the ages, and H. A. Rey either decided on his own to explain it in his book *The Stars: A New Way to See Them*, or perhaps was instructed to publish it in this way, so that it would not be lost (H. A. Rey and his wife Margret had both narrowly escaped Paris ahead of its occupation during the Second World War).

Perhaps it was decided by some of the custodians of this ancient system that it was time for it to be shared more broadly.

Or perhaps H. A. Rey, an accomplished artist as well as an enthusiastic star-gazer, was able to look at the stars and perceive the best outline to match the traditional constellational figures -- and by a process of "great minds think alike," Rey happened to hit upon the same outlines that the ancients had also conceived.

It should be stressed at this point that nowhere in his writings did H. A. Rey suggest that the myths and sacred stories of the world were all based upon the constellations, or that ancient artwork (such as the examples shown above) of gods, goddesses, and heroes could all be shown to correspond to his own outline system. He may have disagreed vehemently with the assertions I make in this book and in my other writings.

But somehow I doubt he would have disagreed.

Celestial Cycles and the Ancient System

The sun rises in the east and sets in the west.

This motion from east to west is not caused by the sun sailing through the sky in a westerly direction, but instead by the earth spinning on its axis towards the east.

As the earth rotates towards the east, the rotation will expose heavenly objects in the east first (from the perspective of an observer on the earth's surface, or flying above it in some sort of flying machine). These heavenly objects or heavenly bodies will be seen to "rise" in the east, out of the horizon, because the earth spinning in that direction will eventually bring them into view to the observer on the ground (or in the air above the earth).

This motion of the earth spinning on its axis creates one of the great heavenly cycles: in this case, the *daily cycle* -- probably the most familiar of all the heavenly cycles. We can hardly go very far in our lives before we become aware of the successive periods of daylight and darkness, which we call *day* and *night*, caused by the fact that as earth rotates on its axis it brings the sun into view to one part of the globe, and then its continued rotation will eventually obscure the sun once again behind the mass of the earth, when the rotation causes the sun to set in the west, creating night-time for the part of the earth which is turned away from the sun and thus is blocked from its rays.

The motion of the sun across the sky (which we can observe as the rotation brings the sun into our view from our location) is thus caused by the earth's turning in an easterly direction, causing the sun to appear to move towards the west (exactly as the forward motion of a car or train will cause houses and billboards and trees outside to appear to move towards the rear of the car or the train).

This very same motion of daily rotation upon its axis by the earth will also cause the motion of the other heavenly bodies we can observe from earth to proceed from east to west in the same manner as the sun's progress. The moon, which can be observed most easily during the night but also during the daytime (depending on its position in relation to the earth and to the location of the observer on the earth who is looking for the moon), also rises in the east and sets in the west.

Likewise for the stars: they begin to come into view as the part of the earth upon which we are located rotates out of the view of the sun (causing the sky to begin to darken), and when they come into view some of them will of course already be high in the sky, but as the earth continues to rotate they too will move towards the west, and we can observe them sinking down into the horizon as the earth continues to spin.

At the same time that some stars are setting in the west, the earth's rotation towards the east will also bring other stars into view, causing them to rise up out of the horizon in the same way that the sun and moon can be seen to do -- all caused by the motion of the earth's rotation upon its axis, which creates this daily cycle of all the heavenly bodies moving from east to west over the course of a single day, each and every day.

The same can also be said for the planets, which from our vantage point on the earth resemble the stars, although planets do not twinkle (any more than the moon twinkles) and although planets have some other characteristics which the visible stars do not share. Despite their differences, planets also can be seen to move towards the west due to earth's rotation on its axis each day.

It should be noted that the two *interior planets* (Venus and Mercury), whose orbits are closer to the sun than is the orbit of our planet, and whose orbits are thus interior to ours, can only be seen if we look generally towards the sun (because their orbits are interior and hence closer to the sun). Therefore, you will usually only be able to observe either their rising or their setting, and not both in the same day. This is because their orbit will put them on "one side or the other" of their close companion the sun, and you will generally only be able to see them when the rim of the horizon is blocking out the sun. In other words, if Mercury is *leading* the sun at one point in its orbit (from the perspective of an observer on earth), we may be able to see it above the eastern horizon in the morning while the sun is still blocked by the curvature of the earth.

However, as the earth's turning continues and brings the sun above the horizon, the blazing light of our day-star will drown out Mercury and cause it to be invisible to observers using their naked eye.

Furthermore, although Mercury will still be ahead of the sun as earth continues to turn and as the sun (and Mercury) sail towards the west, Mercury will remain invisible because the sun in the sky will be filling the heavens with light and render the visible planets and all the stars invisible during the daytime. This will continue to be true even as the sun (led by Mercury) nears the western horizon -- and Mercury will sink down below that horizon before the sun does, if Mercury is in the lead.

Thus, we would only be able to see Mercury prior to sunrise, but not prior to sunset (nor after sunset either) on the same day. When Mercury's orbit brings it around to the other side of the sun and causes Mercury to be "trailing" the sun from the perspective of an observer on earth, then the opposite will be true. Mercury will then be potentially visible only *after* the sun has set in the west, in the early evening, when the western horizon blocks the sun's rays but before the earth's rotation causes close-following Mercury to sink down below the western horizon as well. Obviously, when Mercury is trailing the sun, we would not be able to see it at sunrise in the east, because by the time Mercury clears the horizon, the sun will already be above the horizon, ahead of Mercury – and thus the brightness of the sun's effulgence will drown out the faint planet.

The same can be said for Venus, although the orbit of Venus is exterior to that of Mercury and thus farther from the sun, and so Venus gets "more separation" from the sun on some parts of its orbit (from our perspective on earth) than Mercury can attain. Venus is also larger than Mercury and much brighter (Venus is the brightest of the visible planets), and so Venus can actually be observed sometimes while the sun is still in the sky (although you have to look in the right place, and also be careful not to look directly at the sun, which of course can cause permanent damage to eyesight).

The daily motion, of course, generally brings the same heavenly objects back to their same positions at a rate of once per day. If you are observing the sunrise one morning, then you will be able to observe the sun at nearly the same point again one day later. If you are observing Venus above the western horizon after the sun has set early one evening, then you will see Venus is roughly the same place at roughly the same time one day later. And if you look up in the sky and see the dazzling stars of the constellation Orion one night, then you should see the same constellation at nearly the same location one day later (the next night at the same time).

The assertions in the above paragraph, of course, are only true if we speak very generally -- because there are a great many other cycles going on at the same time, caused by the fact that as the earth is rotating on its axis it is also orbiting the sun, and as the earth is rotating on its axis the moon is also orbiting the earth, and as the earth is rotating on its axis, the other planets are also moving along their own orbits around the sun, introducing other motions into the equation.

Thus, the moon will actually be in a very different spot in the sky one day later, because it moves along its own orbit which creates a *monthly cycle*. The stars will also be at a slightly different place twenty-four hours later than they were the night before, due to the motion of the earth in its orbit around the sun, which is a function of the *annual cycle*. Even the sun's rising and setting times and its exact location of rising and setting along the horizon (and the path of its arc through the sky) will change from day to day due to the fact that earth's axis is oblique (tilted) relative to the plane in which earth orbits the sun -- and this obliquity or tilt introduces *seasonal cycles* which are contained within the annual cycle of earth's orbit around our central star.

And of course the planets have their own cycles, which can become rather complex, although these cycles are not as fast as the moon (because the moon orbiting the earth is so much closer) and so the relative location of the visible planets will not be seen to change as rapidly from one day to the next.

All of the above cycles -- and some others besides -- were imbued with spiritual meaning by the ancient originators of the worldwide system of myth-metaphor. We will have occasion to touch upon some of these cycles, and the layers of meaning with which they were invested, during the course of this book.

For the present, however, we can begin to understand the ancient system of heavenly metaphor which underlies all the world's myths by focusing on the simplest and most familiar of these multiple cycles -- the daily cycle. And, in order to keep from getting too complicated before the foundational concepts are understood, we can begin our exploration of the ancient system by observing what is undoubtedly the most familiar of all the heavenly bodies to the vast majority of men and women living on earth -- our own sun.

The same basic principles and correspondences we unpack in this discussion of the sun and the daily cycle will apply, with some variations and nuances and additional layers of metaphor, for other aspects of the system. This book will focus most on the sun and stars, but you can take the same principles and apply them to other cycles (such as the cycle of the moon).

To put it very plainly and very simply, the ancient myths of humanity ascribe metaphorical meaning to the heavenly motion of rising up out of the eastern horizon, sailing across the celestial expanse, and then sinking down into the western horizon (only to travel back to the eastern horizon and rise up again), and they use this motion to convey a multi-layered body of knowledge regarding the human condition and the world, *the cosmos*, in which we find ourselves during this life.

It is not at all unfamiliar to us to hear that a poetical or metaphorical or allegorical system uses the symbology of day and night, light and darkness, to represent and explore themes regarding life and death and other related pairings such as *this present life in the body* and *the afterlife in a world of spirit*, and thus it will not be surprising to consider the possibility that the ancient system operating in the myths uses the daily cycle to represent and explore these same subjects.

However, what does come as something of a surprise is the fact that the portion of the cycle that is representative of the *other* life, the life outside the body as opposed to this present life, or the realm of spirit rather than materiality is *not* the lower half of the cycle (the night-half, after the sun sinks below the western horizon and before it rises again in the east).

While it might seem that the action of sinking down into the west and disappearing from the heavens would represent going down into *death* and into the land of shades (especially because when the *sun itself* is the heavenly body which is sinking down below the western horizon, the world is plunged into *darkness* until it reappears in the east), following the extensive arguments of Alvin Boyd Kuhn on this subject I am convinced that it is actually the *lower half* of the cycle (following the plunge out of the heavenly realm) which the worldwide myth system employs to represent *life* in this body.[21]

This assertion actually makes sense if we give it some thought. When the sun (or other star) sails across the crystal expanse of the heavenly realms, it can be said to be traveling through the "upper realms" made up of the "higher elements" of fire and air. When the sun (or other star) plunges down into the western horizon (disappearing into the earth -- or, for viewers with a western horizon composed of ocean -- into the sea), it can be said to plunge down into the "lower realms" made up of the "lower elements" of earth and water.

The incarnate life, obviously, consists of the period in which we are present in a physical body, composed of the lower elements of earth and water (or, as the Genesis account tells us, "clay" -- composed of earth and water).

Thus, the plunge down out of the unimpeded heavenly realms to plow through the lower realm of earth and water is used by the ancient myths as the analog for the plunge down into incarnate matter, from the realm of spirit.

The lower realm of night, then, was not seen as the realm of death -- except in the sense that the soul's plunge down out of the realm of spirit was seen as a type of "death." This incarnate life represents the "lower crossing." Being encased in a body thus becomes a kind of spirit-death, and after plowing through this lower underworld passage, the rebirth of the sun (or other star) from the eastern horizon represents the release from incarnation to return to the realm of spirit (in between incarnations).

Alvin Boyd Kuhn explains this metaphorical system by saying:

> [T]his physical body is for the soul the house of death and in its regenerative phase, the house of rebirth. It is the house into which the spirit descends to its partial obscuration in the darkness of the grave of matter, into the night of death, or incarnation, out of which it is to arise in a new birth or resurrection on the opposite side of the cycle. A significant passage from the Book of the Dead recites: "Who cometh forth from the dusk, and whose birth is in the house of death" -- referring to the incarnating soul. In a spiritual sense the soul "dies" on entering the body in incarnation, but has a new birth in it as it later resurrects from it. The body is therefore the

house of death and rebirth, or the place of crucifixion and resurrection.[22]

Elsewhere, Kuhn quotes the ancient Neoplatonic philosopher Plotinus (c. AD 205 - 270) to support his interpretation, saying:

> The body was pictured as the abode of night and gloomy shadows. We have noted Plotinus' statement that in her descent the "soul was precipitated into a darkness profound and repugnant to the intellect," which was obscured by it. The body is "night's dark region" and the soul's "sojourn on earth is thus a dark imprisonment in the body."[23]

Thus, the interplay between light and dark within the daily cycle is used in the ancient myths to convey understanding about the inter-relation of the realms of spirit and of matter -- and the lower half of the cycle represents the arduous journey through matter in *this incarnate life*, although at first glance we might guess that it represents the realm of death.

An example from myth is easy to find. In the mythology of ancient Egypt, there were more than one solar deity. The sun was represented as the falcon-headed sun-god Horus when it soared up from the eastern horizon to cross the sky by day (Horus was often called "Horus of the Two Horizons"). But when the sun plunged down into the west to plow through the underworld through the night passage, it became Osiris, god of the dead, ruler of the lower realm -- the god who was slain.

Alvin Boyd Kuhn elaborates on this mythic symbolism, saying:

> We have seen the sun-god pictured as passing through the dark underworld at night. His voyage is made amid spiritual darkness. The body is the soul's dark prison, grave and tomb. The god is then the sun in the dark

underworld. Therefore it is a light in the darkness. His mission is to bring light into this dark region.[24]

A central theme in all the ancient myths is the depiction of this incarnate life as the arduous passage through the lower realm. Sometimes it is allegorized as the difficult crossing of the stormy sea -- water being one of the two "lower elements" (and our bodies being composed, we are told, of seven-eighths water). Just as the sun and other stars must (metaphorically speaking) burrow through the massy realm of dirt and moisture when they descend from the celestial realms and make contact with the western horizon, so also we ourselves are envisioned as divine souls who have descended from the realm of pure spirit to take on a massy body of clay, toiling through the lower realm during this sojourn in incarnate existence, as we make our way between the western and eastern horizons.

The annual cycle is employed in the myths to expand and elaborate upon this same deep theme -- and this cycle is even more useful for exploring the various stages in the journey, because the fact of the obliquity of the ecliptic further subdivides the year into a lower half when hours of daylight grow shorter than hours of darkness (between fall equinox and spring equinox, with winter solstice in between) and an upper half when hours of daylight again grow longer than hours of darkness (from spring equinox to fall equinox, with summer solstice in between).

Just as in the daily cycle, the "lower half" of the annual cycle (when darkness triumphs over daylight, on the part of the wheel containing at its lowest point the point of winter solstice) represents the arduous crossing of the material realm.

The signs of the zodiac, through which the sun is seen to successively pass as a function of the earth's annual circuit throughout the year, provide further subdivision of the annual cycle, beyond the four great stations of the two solstices and two equinoxes, enabling the myths to deal with the various stages of the soul's journey with much greater precision and subtlety, from the initial plunge into the material realm, in which the soul is immersed in material concerns and forgetful of its true origin, down deeper and deeper until a great "turning-point" is reached, and the soul awakens to its divine nature, and begins to raise itself back to greater and greater integration with its higher nature.

Alvin Boyd Kuhn elaborated with tremendous insight the spiritual significance with which the ancient system of myth imbues this great circuit around the year, from the point of summer solstice down through the plunge into matter at the fall equinox, and then down even deeper to the ultimate nadir at the winter solstice, where the soul reaches its great turning-point. His explication of this cycle, given most clearly in an undated lecture entitled *Easter: The Birthday of the Gods*, and cited in the series *Star Myths of the World and how to interpret them*, is worth reproducing here for its superlative elucidation of the ancient system of spiritual allegory:

> Using solar symbolism and analogues in depicting the divine soul's peregrinations round the cycles of existence, the little sun of radiant spirit in man being the perfect parallel of the sun in the heavens, and exactly copying its movements, the ancient Sages marked the four cardinal "turns" in its progress round the zodiacal year as epochal stages in soul evolution. As all life starts with conception in mind, later to be extruded into physical manifestation, so the soul that is to be the god of a human being is conceived in the divine mind at the station in the zodiac marking the date of June 21. This is at the "top"

of the celestial arc, where mind is most completely detached from matter, meditating in all its "purity."

Then the swing of the movement begins to draw it "downward" to give it the satisfaction of its inherent yearning for the Maya of experience which alone can bring its latent capabilities for evolution of consciousness to manifestation. Descending then from June it reaches on September 21 the point where it crosses the line of separation between spirit and matter, the great Egyptian symbolic line of the "horizon," and becomes incarnated in material body. Conceived in the aura of Infinite Mind in June, it enters the realm of mortal flesh in September. It is born then as the soul of a human; but at first and for a long period it lies like a seed in the ground before germination, inert, unawakened, dormant, in the relative sense of the word, "dead." This is the young god lying in the manger, asleep in his cradle of the body, or as in the Jonah-fish allegory and the story of Jesus in the boat in the storm on the lake, asleep in the "hold" of the "ship" of life, with the tempest of the body's elemental passions raging all about him. He must be awakened, arise, exert himself and use his divine powers to still the storm, for the elements in the end will obey his mighty will.

Once in the body, the soul power is weighed in the scales of the balance, for the line of the border of the sign of Libra, the Scales, runs across the September equinoctial station. For soul is now equilibrated with body and out of this balance come all the manifestations of the powers and faculties of consciousness. It is soul's immersion in body and its equilibration with it that brings consciousness to function.

Then on past September, like any seed sown in the soil, the soul entity sinks its roots deeper and deeper into matter, for at its later stages of growth it must be able to utilize the energy of matter's atomic force to effectuate its ends for its own spiritual aggrandizement. It is itself to be lifted up to heights of cosmic consciousness, but no more than an oak can exalt its majestic form to highest reaches without the dynamic energization received from the earth at its feet can soul rise up above body without drawing forth the strength of body's dynamo of power. Down, down it descends then through the October, November and December path of the sun, until it stands at the nadir of its descent on December 21.

Here it has reached the turning-point, at which the energies that were stored potentially in it in seed form will feel the first touch of quickening power and will begin to stir into activity. At the winter

> solstice of the cycle the process of involution of spirit into matter comes to a stand-still -- just what the solstice means in relation to the sun -- and while apparently stationary in its deep lodgment in matter, like moving water locked up in winter's ice, it is slowly making the turn as on a pivot from outward and downward direction to movement first tangential, then more directly upward to its high point in spirit home.
>
> So the winter solstice signalizes the end of "death" and the rebirth of life in a new generation. It therefore was inevitably named as the time of the "birth of the Divine Sun" in man; the Christ-mas, the birthday of the Messianic child of spirit. The incipient resurgence of the new growth, now based on and fructified by roots struck deep in matter, begins at this "turn of the year," as the Old Testament phrases it, and goes on with increasing vigor as, like the lengthening days of late winter, the sun-power of the spiritual light bestirs into activity the latent capabilities of life and consciousness, and the hidden beauty of the spirit breaks through the confining soil of body and stands out in the fulness of its divine expression on the morn of March 21. This brings the soul in a burst of glorious light out of the tomb of fleshly "death," giving it verily its "resurrection from the dead." It then has consummated its cycle's work by bursting through the gates of death and of hell, and marches in triumph upward to become lord of life in higher spheres of the cosmos. No longer is it to be a denizen of lower worlds, a prisoner chained in body's dungeon pit, a soul nailed on matter's cross. It has conquered mortal decay and rises on wings of ecstasy into the freedom of eternal life.[25]

This explication of the esoteric spiritual significance of the great annual cycle demarcated by the four stations of the two solstices and two equinoxes by Alvin Boyd Kuhn is absolutely indispensible for the understanding of the ancient system at the base of all the world's myths, scriptures and sacred traditions. And the more we study the myths, the more we realize that Kuhn's assertions are borne out in the message which they are trying to convey to our understanding.

However, despite his mention of the Scales of Libra in the above-cited passage, which Kuhn inserts into the discussion of the point of the September equinox (where spirit and matter are "weighed in the balance" and found to be equal, just at the point where day and night find equal balance and just at the point at which spirit incarnates into a human being possessed of both a physical and a spiritual nature), Alvin Boyd Kuhn does not in his writing appear to be aware of the very close correspondence between the personalities and events we see depicted in the myths, and *specific constellations* in the night sky.

Had he done so, Kuhn would doubtless have instantly perceived that this correspondence imbues the myths with a complete code or language with which to describe *with great precision* the various stages of the great cycle he describes with such eloquence and insight in the extended passage cited above.

For the constellations of the night sky give us a much more precise demarcation or subdivision of the great circle, beyond the four great stations of the solstices and equinoxes, just as a clock-dial or watch-face which contains *all twelve* numbers allows us to tell time with greater precision than does a more "modernistic" watch dial containing only four dots located at the 12 o'clock, 3 o'clock, 6 o'clock, and 9 o'clock positions. And a watch-dial containing small "ticks" between each of the twelve numbers is even more precise, allowing us to see exactly how many minutes past the hour it is at any given moment. Try determining when it is exactly 13 minutes past five o'clock on a watch without "ticks" between the twelve numbers, enabling you to count the precise minutes – let alone trying to confidently tell when it is exactly 5:13 on a watch that only contains four dots at the four cardinal points of the circle!

The twelve constellations of the zodiac appear in coded form throughout the ancient myths and sacred stories -- especially the constellations that populate the "lower half" of the cycle, which corresponds to this incarnate life, when we are making our way across the difficult "underworld journey." Below is a diagram of the great wheel of the year, with the zodiac signs arranged in clockwise manner, such that the sun's progress through the year is depicted in a clockwise direction (some zodiac wheels arrange them in the opposite direction, although I prefer this way):

Readers familiar with my other writings will recognize this diagram as one I use quite frequently to show the great circle of the annual cycle. The circle is arranged such that the upper and lower halves of the year can be easily divided by the horizontal line which I have added between the two large "X" symbols which I have added on either side.

These two "X's" mark the two equinox points, where the year "crosses over" from having daylight dominating on the upper half (when hours of daylight are longer than hours of darkness) to having darkness dominate on the lower half (when hours of night-time become longer than hours of daylight). The X on the right side of the wheel as we face the page marks the point of fall equinox (for the northern hemisphere), when we "cross down" from the upper half of the year to the lower half, on the way to the winter solstice (which is located at the 6 o'clock position at the bottom of the wheel), and the X on the left as we face it marks the point of spring equinox.

The ancient myths do not restrict themselves to the twelve zodiac constellations alone, however. The zodiac band represents that circle of stars arranged along the plane of the ecliptic, the plane formed by earth's orbit around the sun. If you imagine the earth orbiting the sun and "shrink it down" in your mind to a size where you can stand outside the orbit and observe it as if it is taking place within one of the rooms in your house, and then imagine a laser-pointer fastened onto the earth and pointing its beam directly at the sun all the time as earth makes its way all the way around the orbit, and then imagine that laser-pointer's beam continuing through the sun to the far wall, then you can perhaps envision that the laser beam will trace out a line around all four walls as the earth with its attached pointer makes its way in a full circuit around the sun.

The line along the wall would trace out the path of the zodiac -- and an imaginary plate of glass fitted within the four walls at the level of the line (or an elliptical plate of glass within the orbit of the earth around the sun) would be the plane of the ecliptic.

So the zodiac constellations are the figures formed by the stars which have part of their outline crossing that "line" burned onto the wall by the laser-pointer in our imaginary mental model described above. But there are also constellations above and below this line (on all four walls of the room, and on the floor and the ceiling besides). These constellations also have spiritual significance in the great system underlying the myths of the world -- especially those constellations "above" the line and up to the ceiling of the room, which is to say "north" of the line, if the north pole in our mental model is facing the ceiling, perhaps because the ancient system does seem to be "northern-hemisphere-centric" in many ways (even when it shows up in Africa and Australia and other parts of the world south of the equator). We will have occasion to mention some of these constellations, and their spiritual significance as employed in the myths, later in the book.

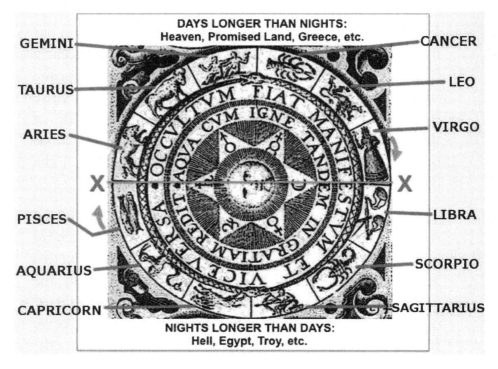

There is also an extremely important celestial figure in our night-time sky which is not a constellation, but which rather represents the bright band of light created by the massive number of stars in our galaxy itself, and which is commonly known as the Milky Way band (our galaxy has in fact been given its name after the name for this celestial feature in our night sky – the Milky Way galaxy).

The galaxy is shaped like a giant disc, with a bulge in the center somewhat like a fried egg with a bulging yolk at the middle. When we look out into the night sky, the brightness of the Milky Way is created by the plane of this great disc, within which we (and our solar system) are actually situated. There is actually a bulge in the Milky Way which represents the galactic core – the "yolk of the egg" so to speak.

In our mental model of the earth's orbit inside a room in our home, the Milky Way could be spray-painted in a great circle running up one wall, across the ceiling, and down the opposite wall and across the floor. It would intersect the zodiac circle that runs around the room at two points, on opposite sides of the room, and it would intersect it between the constellations of Scorpio and Sagittarius on one side and between the constellations of Gemini and Cancer on the other side.

The Milky Way is kind of like a huge wedding ring within which we on earth are orbiting around the sun, and when we look out into space we see the interior surface of that ring (it's actually a huge flat disc that we are inside, but it looks like the inner surface of a ring as we look out at it from our vantage point on earth, as if we are in the "eye" of a hurricane). As we turn on our axis each day (the daily rotation, which creates the daily cycle discussed earlier), the ring seems to "flip" over us in the sky (due to the spinning of earth on its axis).

It makes one complete flip each day as we spin, although we only get to see one side of the interior of that ring on any given night, since the other half of the flip is generally obscured by the sun's light during the daytime. We also generally get to see one portion of the inner surface of that giant wedding ring during each half of the year.

We only see the Milky Way at night, of course, so when we are on the side of the orbit when we face away from the sun and see Scorpio and Sagittarius, we see one half of the inner surface of that ring, and when we are on the other side of the orbit and see Gemini and Cancer when we are turned away from the sun, then we see the other half of the inner surface of the ring.

The brightest half of the Milky Way band is the part we see rising between Scorpio and Sagittarius, and this portion of the Milky Way also contains the great bulge of the Galactic Core, which makes a noticeable widening in the band in that region (this is the "yolk of the egg" of the relatively flat disc of the galaxy).

The Milky Way band is extremely significant in the spiritual system of celestial allegory informing the world's ancient myths. The bright portion of the galaxy rising between Sagittarius and Scorpio corresponds closely to the point of winter solstice, which is the great "turning-point" described in the extended quotation from Alvin Boyd Kuhn included in the discussion above. This is also the portion of the galaxy containing the Galactic Core, which not only features a prominent bulge but also a dark channel known as the Great Rift, which in some sacred traditions we know to have been associated with a birth canal (among the Maya, for certain). It is extremely interesting that this "birth canal" corresponds closely to the great "turning-point" in the annual cycle, where we experience a "second birth" -- a birth of awakening to the spiritual nature, after our initial plunge into incarnation and into greater and greater entanglement with the lower nature of the physical body.

My analysis of the myths suggests that the orientation of the Milky Way band, which is roughly perpendicular to the parade of the zodiac constellations and which thus can be seen to rise vertically out of the horizon (sometimes -- remember that it is flipping over and over our heads as if it is a great ring and we are inside the flipping ring), translates in myths to its often being used as a pathway which represents *the elevation of the spirit*. We can see this very clearly, I believe, in the Odyssey of ancient Greece, which I have explored at some length in *Star Myths of the World, Volume Two* (focused almost entirely on the myths of ancient Greece).

The Milky Way thus functions, I believe, as a symbol of the elevation of our consciousness, or the raising of our internal energy (which is termed *chi, hei,* and *prana,* among other names, by various cultures which have retained a connection to the ancient wisdom). It is also related to the raising of the kundalini energy along the channels through the chakra points as described in the traditions of India and other surrounding cultures.

More generally, when we see the myths describing motion *up* or *down* the band of the Milky Way, they seem to be discussing the motion towards greater awareness of and integration with our spiritual nature and in fact with our divine nature, with our higher self, and with our internal connection to the Infinite.

As mentioned before, there are many other heavenly cycles which feature in the ancient Star Myths of the world, and we will have occasion to examine the significance of some of them in the discussion which follows. Some of most important of these, from a mythological perspective, include the cycle of the *moon*, and the great ages-long cycle of *precession* (sometimes called "the precession of the equinoxes," and sometimes referred to as the Great Year, or the Platonic Year). The precessional cycle also appears to be related to the cycles of the Yugas described in the sacred Sanskrit texts of ancient India.

All of these cycles, and their spiritual significance, however, can basically be understood to display a similar pattern to that described in this chapter, and to deal with the endless interplay between matter and spirit, and between the lower nature and the higher nature, whether playing out in the waxing and waning of the moon every month, or the lengthening and shortening of the days throughout the year, or even the inexorable and ages-long grinding of the gears of precession.

All of them are used by this ancient worldwide system to convey a profound and multi-layered body of knowledge, knowledge which (according to the message contained in the ancient myths) we must understand in order to gain an understanding of the nature of the universe in which we find ourselves, and the nature of the human condition in this incarnate life: who we are, why we are here, and what we are supposed to be doing in this simultaneously spiritual-and-material cosmos.

Clothing the Heavenly Cycles in Myth

As we have already seen from some of the discussion in the previous chapter, the heavenly cycles created by the circling of the earth around the sun, the moon around the earth, the motions of the planets, and even (potentially) the interaction of our sun with a possible binary partner,[26] are employed by the ancient worldwide system to "give form" to deep and nearly-unspeakable concepts – concepts dealing with invisible and spiritual matters, which are difficult to convey directly and which have so many layers of meaning that they almost stagger the mind.

Therefore, the ancient wisdom cloaks its profound teachings in a robe of metaphor, turning the interplay of light and dark throughout the course of the year (for example) into a massive battle between opposing armies, or the descent of the sun towards the solstice as the shearing-off of the radiant locks of hair of a mighty hero as he sleeps "upon the knees" of Delilah.

I believe, following the arguments of the esoterist R. A. Schwaller de Lubicz (1887 - 1961), that the system cloaks its truths in metaphor not in order to *hide* them but rather in order to best *convey* them to the understanding of our heart.[27] By arraying these truths in garments familiar to our experience here in the material realm, the Invisible Realm can be said to "condescend" or stoop to our level, in order to convey the divine world in the terms and images and activities of the ordinary world.

And indeed, a large part of the message does seem to involve the understanding that this apparently ordinary, apparently material world which we think we inhabit is infused and interwoven at every single point with the divine realm, the realm of spirit. The world we inhabit is not so ordinary at all – and neither are we ourselves – and is invisibly but inseparably connected to the realm of "non-ordinary" reality and experience.

The "thickness" (so to speak) of this veil of metaphor or esoteric allegory can vary significantly between one ancient tradition and the next – or even within the same tradition.

Sometimes we see the heavenly cycles portrayed with rich layers of mythical figure, particularly notable (for instance) in the multitudinous divine and semi-divine figures taking part in the adventures described in the ancient Sanskrit epics of the Ramayana and the Mahabharata, in which we find entire pantheons of gods as well as well-developed heroes and heroines with minutely-described accouterments and well-rounded personalities, and in which the struggles between the various protagonists and antagonists are framed in narrative action that relate directly to events familiar to human existence, such as battles, courtships, weddings, arguments, contests of skill, and games of dice.

Other examples include most (but not all) of the stories in the so-called Old and New Testaments of the Bible, as well as most of the myths of ancient Greece or northern Europe or the Americas and the Pacific and the continents of Africa and Australia and much of Asia. The layer of metaphor is "thickest" (according to this description) when we can barely perceive that the stories are not intended to be seen as describing literal events, albeit events which occasionally seem miraculous (or pertain to a bygone age, when miracles and revelations and the visit of deities must have been more frequent).

In these stories, even if miraculous, we can at least envision what is going on and relate it to experiences and situations which we ourselves have either faced or at least observed in our "ordinary" lives.

But sometimes the veil of metaphor which is thrown over the mysterious workings of the heavenly cycles (which themselves represent the mysterious and numinous Invisible or Infinite Realm) is much "thinner," and much less concerned with cloaking the heavenly matters in forms or episodes which we can recognize from our own daily experiences. In these cases, metaphor is still present, but (for whatever reason) the attempt to make the metaphors fit in with a coherent narrative is much less pronounced.

A good example of this, familiar to many readers from western traditions where the scriptures of the Old and New Testament have exerted strong influence for many centuries, can be found in "apocalyptic" literature or traditions, such as we find in the book of the Revelation of John found at the end of the canonical New Testament.

There, the metaphors are still celestial in nature (many of them having to do with the great shifting of the precessional ages, although many of them also relate to the very same specific constellations which form the basis for the events narrated in the gospel accounts as well as for the episodes in epics such as the Mahabharata or the Iliad and the Odyssey), but now they take on weird and unnatural forms. No attempt is made to frame them in a narrative that would enable us to mistake them for something that *could* happen in ordinary human life.

For example, in the vision described in Revelation chapter 1, the narrator describes seeing seven golden candlesticks, and in the midst of them one "like unto the Son of man" with feet like brass, seven stars in his right hand, and a sharp two-edged sword coming out of his mouth.

Later, in the ninth chapter, a star falls to earth and is given "the key to the bottomless pit," and when the bottomless pit is opened, smoke arises from it, followed by beings described first as locusts but also having "power as the scorpions of the earth have power," and later as having shapes that are "like unto horses prepared for battle" as well as faces like the faces of men, hair like the hair of women, teeth like the teeth of lions, and crowns of gold. After all of these descriptions, it is hard to imagine exactly what part of these beings actually warrant their description as "locusts" -- but in fact all of the features described are celestial in nature and relate to specific constellations and their outlines and features and location in the sky, as I have detailed in other volumes, following arguments presented in *Hamlet's Mill* (the rising smoke is the wide and bright section of the Milky Way which rises between Scorpio and Sagittarius, and most of the features described in Revelation 9 relate to this region of the sky).[28]

Many other examples of this "less naturalistic" style of myth-metaphor can be cited from around the world. For example, many of the texts that decorate the temples and pyramids and reliefs and sculptures of ancient Egypt refer to events taking place in the divine or Invisible Realm in a manner that seems to have more in common with the style we see in the book of Revelation (less "naturalistic" and more "bizarre" or unworldly) than with the myths of ancient Greece or the narratives of the Old and New Testament (in which the symbols are fitted into a framework that may include miraculous events but which more resemble events we might see in "ordinary reality" rather than visions we might see in a psychedelic vision or a dream).

Similarly, the events related in the Kojiki of ancient Japan do follow a narrative, but the level of "naturalistic" metaphor is sometimes more and sometimes less naturalistic: many of the events described are completely non-naturalistic in nature and again resemble what we see described in some of the chapters of Apocalyptic books such as John's Revelation. The narrative is not always smooth, and the actions are described without much context or motive being layered around them to try to make the story more coherent.

Again, just as in all other myths, I believe the visions and events described in the Kojiki (or in the surviving texts from ancient Egypt) can be traced to constellations and heavenly cycles, and just as in the other myths, I believe they are intended to depict and convey deep truths regarding the Invisible Realm which touches this Visible Realm at every point (and from which, indeed, the Visible Realm originates and proceeds). But, because of their "less naturalistic" nature, we may find these types of mythical narratives to be somewhat less accessible or approachable than (for instance) the stories of ancient Greece or of ancient India.

If the Invisible Realm is an infinite void, and if the metaphors with which the ancient myths clothe this Invisible Realm are described as a veil or a gown or a robe, then we might say that in some of the mythical traditions, the veil is so thin that we find ourselves staring more directly into the unadulterated void, with less to mediate or soften the experience, and with less that is familiar or recognizable offered to us as a sort of "handhold" to which we can cling as we gaze into that vast abyss.

Before leaving this brief discussion of the "thickness of the metaphor" and getting into some exploration of the outlines of the metaphorical garments themselves, we should also acknowledge that in some manifestations of the ancient wisdom, the metaphorical layer is so thin and translucent that it seems almost nonexistent.

In the Tao Te Ching of ancient China (sometimes spelled Dao De Jing), for example, there is almost no narrative at all which could be said to resemble the elaborate narratives of ancient India or ancient Greece or the Polynesian narratives of Maui. Nevertheless, as I have discussed previously (such as in *Star Myths of the World, Volume One*), within the Tao Te Ching we still find references to the arising and departing of the "ten thousand things" (or the "myriad creatures," as Professor Victor H. Mair translates them), which I believe to be describing the same interplay between the Invisible and Visible Realms which is associated elsewhere in myth with metaphors such as the struggle between the Achaeans and the Trojans upon the plains of Ilium, or the battle between the Pandavas and the Kauravas on the plains of Kurukshetra.

Interestingly enough, the very expression "ten thousand things" is made up of the Chinese characters 萬物, the first of which means "ten thousand" or "a vast number" but which incorporates a symbol which also means "a scorpion." Whether or not this ancient phrase reveals a connection to the constellation of the Scorpion, which is indeed found near the "bottom of the annual cycle" and the "turning point" described in the extended quotation from Alvin Boyd Kuhn cited in the previous chapter, as well as near the Great Rift feature in the Milky Way, which some sacred traditions describe as a celestial "birth canal," can be debated -- but it *is* an intriguing possibility.

The Chinese character for the number 10,000 actually incorporates the radical for "insect," which is 虫. Most characters for words having to do with insects, bugs, worms, crustaceans, and other invertebrates incorporate this insect radical in the character. While it may be argued that it is not so surprising to find that the character for "a myriad" or "ten thousand" would incorporate the "insect" radical (since many insects do, after all, seem to swarm in their myriads), it may be considered somewhat more unusual to select the specific character for "scorpion" to correspond to a "myriad" or to the number "ten thousand."

The connection seems somewhat difficult to explain -- except for the fact that ancient texts such as Tao Te Ching reveal the use of this character to express the idea of "the myriad things" unfolding into existence (from the realm of pure potential, the implicate realm, or the Infinite Realm), and we know from other myths around the world that the zodiac wheel was used in other ancient traditions to express the same idea, with the point of "rebirth" being located between Scorpio and Sagittarius on the wheel.

Thus, I believe that even in ancient traditions in which the "ornateness" of the metaphor is very subdued (such as the Tao Te Ching, which is almost the opposite of the ancient Sanskrit myths of India, for example), the same ancient system of celestial metaphor can be detected.

In fact, as I have also written about elsewhere (such as in *Star Myths of the World, Volume One*), the traditions surrounding Lao Tzu himself (or Laozi, or Lou Dzai to use the Cantonese pronunciation) contain clear references to the heavenly cycles and give evidence of being built upon the same worldwide system of celestial metaphor we find in other cultures very far removed from ancient China.[29]

The same can also be said of the ancient Buddhist traditions, which are often described as being anti-mythical in nature and as having arisen in part as a reaction to the ornate and elaborate myths and stories found in the Hindu tradition, with their countless gods and goddesses and celestial beings. But once again, even the traditions surrounding the supposedly historical events in the life of the Buddha can be shown to be built upon the same celestial system which forms the basis for the more multifarious personalities and stories-within-stories of the myths of ancient Hinduism.

The more we study the many different ancient traditions from around the world from a perspective of their common celestial underpinnings, the more we will begin to see some of the commonalities in the framework of the ancient structure -- commonalities which remain fairly consistent, regardless of the level of "decoration" carved into the pillars and beams of the structure itself.

The decoration may be highly elaborate and mythical, as with the countless gods and goddesses described in the Sanskrit epics of Mahabharata and Ramayana (the first of which alone is about 7.2 times longer than both the Iliad and the Odyssey *combined*), or the decoration may be very simple, as with the Tao Te Ching. But in each case, the same ancient metaphorical system is used to convey a very similar vision of the Invisible World which permeates every aspect of the Visible World (and to which we also have an inner connection).

With that understanding, let's briefly explore a few examples of the way in which the ancient worldwide system can clothe the awesome mechanics of the heavenly cycles in a garment of metaphor.

In the previous chapter, we saw the extended quotation from Alvin Boyd Kuhn arguing that the annual cycle -- delineated by the four points of summer solstice, fall equinox, winter solstice, and spring equinox -- is used in ancient myth to envision and embody the interplay of spirit and matter, and specifically the soul's plunge into the material realm, its initial entanglement in things material, and its eventual transcendence of the material realm . . . back to the realm from whence it originally came.

Alvin Boyd Kuhn himself perceived that the Osiris cycle of myth from ancient Egypt was a mythical embodiment of this principle of spirit's manifestation in matter and ultimate transcendence of that material condition. In *Lost Light*, he writes:

> The savior is not nailed *on* the tree; he *is* the tree. He unites in himself the horizontal human-animal and the upright divine. And the tree becomes alive; from dead state it flowers out in full leaf. The leaf is the sign of life in a tree. The Egyptians in the autumn threw down the Tat cross, and at the solstice or the equinox of spring, erected it again. The two positions made the cross. The Tat is the backbone of Osiris, the sign of eternal stability.[30]

If we unpack the deep analysis that Kuhn presents in the above passage, we see the outlines of a fundamental metaphor which flows through many myths from around the world.

The "backbone of Osiris," which Kuhn refers to as the "Tat" but which in more recent translations is usually rendered the "Djed," is "thrown down" at autumn equinox, when the divine spirit-nature plunges into incarnate matter. In the Egyptian myth, this is associated with the slaying of the god: the death of Osiris. The divine nature takes on a mortal nature: the god is slain and imprisoned in the "underworld" – that lower realm of matter, represented in the ancient myth-system by the lower half of the Great Wheel of the year, from the autumnal equinox down to the winter solstice, and all the way across the "lower months" until we reach spring equinox.

As Alvin Boyd Kuhn explains in the above quotation, the god Osiris is laid out *horizontally* upon his bier in the realm of the dead, corresponding to the imprisonment of the divine soul within an animal, material body (just as animals tend to walk around horizontally, on all fours).

Nevertheless, this sojourn in the lower realms of matter is only temporary: the backbone of Osiris will be *raised back up* to an exalted and upright position again, representative of the re-awakening of the divine nature, and the restoration of the divine nature (which has been sleeping under the stupefying influence of the material realm).

The "casting down" of the divine nature (pictured in Egyptian myth as the slaying of Osiris and the casting down of the Djed column, or "backbone of Osiris") is thus associated with the "horizontal line" between the two equinoxes, which divides the "lower half" of the year from the "upper half" -- and demarks the realm of matter, and incarnation in this animal body of flesh.

The "raising back up" of the same divine nature -- the reconnection with the Infinite, the recognition that we ourselves and everyone we meet is more than just the physical and material nature -- is represented by the "vertical line" running from the lowest point of winter solstice, located between Sagittarius and Capricorn on the preceding zodiac diagram (the "turning point," according to Alvin Boyd Kuhn), and the highest point of summer solstice at the top of the chart, located between Gemini and Cancer.

In the myth of Isis and Osiris, depicted in various wall paintings and reliefs in ancient Egypt, and later described by the esoteric initiate Plutarch, the slain Osiris is encased in a coffin, which later becomes enclosed within a tree (usually a tamarisk tree), which then is chopped down and used to make a pillar within the palace of the king and queen of Byblos.

The goddess Isis, searching all over the world for her beloved Osiris, finally locates the pillar, and requests its return. The panel on the following page shows a temple relief from the Great Temple at Abydos, in which the goddess Isis is receiving the pillar containing her consort Osiris from the king of Byblos. The form of the pillar, with its four "vertebral" segments, clearly indicates that this is a Djed (or Tat) pillar -- representative of the "backbone of Osiris" which had been "cast down" upon the death of Osiris, but which will now be restored.

The scenes of the Virgin Mary receiving the body of Jesus from the Cross parallel very closely the imagery and themes of the recovery of the Djed of Osiris by the goddess Isis.

But this grand pattern of being "cast down" into pain, humiliation, and death in the material realm, only to be "raised up" again into transcendence (and the uplifting not only of the divine spiritual nature but the material nature as well) finds echoes in myths which go far beyond the obvious echoes in the gospel stories of the crucifixion and the resurrection.

We find the "chopping down" of a great pillar or tree, invested with great import and significance, depicted again and again in the ancient mythology and sacred traditions of the world.

We see the chopping down of a heavenly cedar-tree in the Gilgamesh series of myth from ancient Sumer and Babylon (indeed, the chopping down of this great cedar forms one of the most central episodes in the entire Gilgamesh series).

We see the chopping down (or the threatened chopping down) of the great post of the bed of Odysseus (which is fashioned from a living tree-trunk) forming the crucial tension at the very climax of the Odyssey of ancient Greece.

We see the splitting of the great World-Tree, Yggdrasil, at the terrible day of Ragnarök, at the violent end of the world described in the ancient Norse myths.

And there are many more from around the world, all of which have strong parallels to the mythical pattern just described.

Without delving into all the specific details contained in each of the above examples, we can see that they all "clothe" the heavenly motions in layers of mythical metaphor -- metaphors which differ on the surface in their specific details (the details in the story of Gilgamesh and Enkidu on their quest to chop down the tallest tree in the sacred cedar forest, for example, are superficially very different from the details surrounding the death and resurrection of Osiris, or of the Christ, or the details surrounding the confrontation between Penelope and Odysseus and the great tree-post of the long-suffering couple's marriage bed), but which all embody the same great mythical pattern.

It is a pattern which, as intimated above in some of the quotations from Alvin Boyd Kuhn, involves the tension and interplay between spirit and matter -- between the manifest realm of the material, and the invisible realm of pure potential.

I would submit to the reader that the passages in the Tao Te Ching describing the appearing and the disappearance of the "myriad things" (the 萬物 or "*maan mat*") have to do with the very same endless folding and unfolding between the divine invisible realm and the visible material realm: casting down, raising up.

This brief examination introduces us to some of the ways that the ancient system of myth appears to "drape" the celestial cycles with sheets of metaphor. Sometimes the "sheets" are thicker and more elaborately woven or embroidered -- and sometimes they are rather gauzy and transparent, barely cloaking the cosmic cycling at all (I would argue that the Tao Te Ching sheathes the cycles in a much more gauzy and diaphanous sheet of metaphor, for example, than we find in the events depicted in the Iliad or the Mahabharata).

And yet, through them all -- once we begin to become more familiar with the system -- we can perceive the outline of the same underlying structure. It is as if different sheets are being thrown over the same group of furniture: no matter how much the sheets themselves differ from one another in their patterns and decorations and thicknesses (some of them being transparent gauzy films, and others resembling thick rugs or blankets), the furniture below does not change, and so the general outline of the blanketed group will always be roughly the same.

If we take the above metaphor one step further, and imagine that the "group of furniture" over which the different "sheets" are being draped is in fact an *invisible* group of furniture, then we understand that we cannot see the group of furniture at all unless we drape something over it (without the sheet over the top of this furniture, we might stumble over it and bruise our shins). It is the sheet of metaphor which enables us to perceive what is underneath at all -- without the drapery of the metaphor, we would not be able to see it (because it is invisible). But with the draping of myth, we can make it out, and perceive the general outline of the "invisible furniture" that is underneath -- and it doesn't really matter whether the draping is one color or another, whether it is thick and rug-like or thin and filmy.

And this is what, I am convinced, the ancient myths are doing (at least on one level): they are showing us, through celestial metaphor, truths about an Invisible Realm which we otherwise would be unable to perceive, or to grasp.

Casting Down and Raising Up: Orion & the Djed

Let's continue to examine some of the patterns mentioned in the previous chapter, patterns of allegory or metaphor which the ancient myths clothe or cloak the great heavenly cycles, in order to convey to our understanding profound truths which might otherwise be difficult or impossible for us to perceive and to grasp.

The previous chapter introduced some of the evidence arguing that the myth in which Osiris of ancient Egypt is described as being slain and sent to the underworld embodies in story-form a pattern which relates to the endless interplay between light and darkness, daylight and night-time, summer months and winter months, created by the annual cycle of our earth's orbit around the sun (and by the tilt of our axis as we orbit, which causes the path of the sun's journey to be higher across the sky during half of the year -- and the rays of sunlight upon the earth to be more direct -- and lower across the sky during the other half of the year).

We have seen that this endless interplay between the two halves of the year -- one in which daylight dominates over night-time during the summer months (when hours of daylight are longer than hours of darkness), and one in which night-time dominates over daylight during the winter months (when hours of darkness are longer than hours of daylight -- was embodied in many of the world's Star Myths as an epic battle between two opposing forces, such as the Achaeans versus the Trojans in the Iliad of ancient Greece, or the Pandavas versus the Kauravas in the Mahabharata of ancient India.

But we also in the previous chapter began to explore the evidence arguing that this same endless interplay was embodied in ancient myth in other guises as well, such as in the cycle of myths from ancient Egypt involving the "casting down" of the Djed column (the "backbone of Osiris") into a horizontal state (when the god is laid out upon a funeral bed), and the subsequent "raising back up" of the same divine figure, when Osiris is restored through the power of the goddess Isis and by the power of Horus, the son of Isis and Osiris.

Interestingly enough, we know from ancient sources that the figure of Osiris himself was associated by the Egyptians with the dazzling constellation of Orion in the night sky.

For example, in the "diagonal calendars" painted on the lids of Egyptian coffins during the Middle Kingdom, figures associated with specific "decanal stars" would be depicted moving across the rectangular calendar, each "decan" corresponding to a star whose heliacal rising was associated with a ten-day period, and succeeded by another decanal star which would become visible just prior to sunrise during the succeeding ten-day period (this process is explained in *Star Myths of the World, Volume Three*).[31]

While we do not know the identity of all of the stars associated with the figures on Egyptian diagonal calendars or star-clocks, we do have substantial evidence which argues that the star Sirius was the first and most important of them all, and was associated with the figure of the goddess Isis (who appears in the diagonal calendars) — and that the stars of Orion functioned as decans in the ten-day periods preceding the appearance of Sirius.[32]

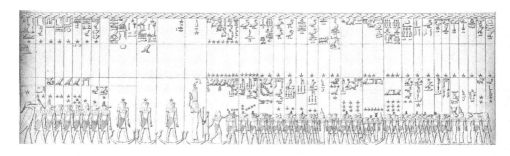

Above: Rectangular star-clock from the tomb of King Seti I (thought to have reigned from 1290 BC until 1279 BC). Goddess Isis is the tallest of the decanal figures, and to the right as we face the image we see a figure associated with the constellation Orion.

A similar rectangular star-calendar from the tomb of the architect Senenmut (who lived during the reign of Queen Hatshepsut, which was thought to have been between 1478 BC and 1458 BC), shows the figures of Isis and the striding constellation of Orion, and in this case the three distinctive "belt-stars" of the constellation are depicted over the figure's head, so that the identification is unmistakable:

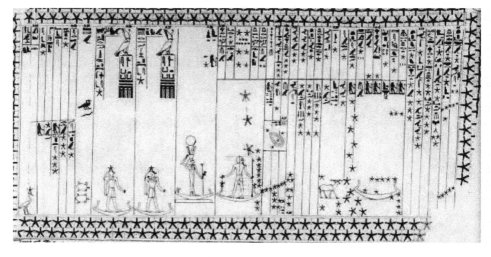

Above: Portion of the rectangular star-clock from the tomb of Senenmut. Once again, Isis is the tallest, and preceding her (to the right as we face the image) is a figure associated with Orion. Note the three "belt-stars" in his rectangular space.

The name that the ancient Egyptians gave to the constellation which we call Orion was *Sahu*, or *Sah*, a god associated with the constellation. It is fairly well accepted among scholars that the god Osiris (Egyptian name was closer to *Asar* — our more familiar *Osiris* is a Greek version of the name) was closely associated with Sahu. For example, Otto Neugebauer and Richard A. Parker declare in *Egyptian Astronomical Texts* (1960) that: "In the Pyramid Texts [the constellation] Sahu is identified with Osiris, which fits well with its depiction as a human on the coffins and ceilings."[33]

Further evidence for the identification of a close association between the god Osiris and the starry outline of the constellation Orion can be found in the numerous depictions of the god lying upon his funeral bier or upon a lion-couch, while still displaying the distinctive striding posture of the Orion constellation! Below is an example of one such image, in which the god is horizontal upon the couch, and yet depicted as if striding forward:

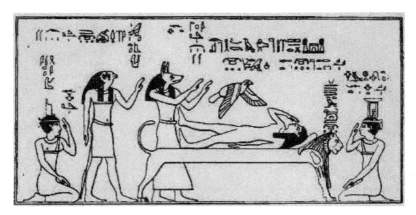

Above: Illustration of the scene of Isis in the form of a falcon above Osiris on a lion-couch, published by E. A. Wallis Budge in 1911 and based upon a temple-relief at Dendera.

From all of this, we can now state that Osiris is the god who is cast down and slain, but who later returns after he is raised-up again, and that this god, in ancient Egypt at least, was associated with the constellation Orion.

The heavenly cycles can be seen to create this motion of "casting down" and "raising up" on various levels. First, there is the daily motion caused by earth's rotation upon its axis. This motion, as we have already discussed, causes all the heavenly bodies (sun, moon, stars, and planets) to rise in the east and set in the west. This motion, with respect to the constellation Orion, causes the outline of the striding figure to appear horizontal when rising in the east or setting in the west, but vertical when crossing the center of its arc across the sky (above the southern horizon, for viewers in the temperate latitudes of the northern hemisphere, and above the northern horizon for viewers in the temperate latitudes of the southern hemisphere).

Above: Composite image showing Orion in three positions – first rising in the east (left), then crossing top-of-arc (center), finally sinking down into the west.

Note that in the above image, the horizon "curves," to simulate the fact that in real life when standing outdoors and facing south, the east will actually be ninety degrees left and west will be ninety degrees right. Therefore, as you look at Orion's position when rising from the east and setting in the west, imagine the horizon being horizontal as it would be if you were outside -- which means that the outline of Orion would be more horizontal as well. This helps to visualize the way in which the daily cycle enacts the motion of Osiris being "cast down" and "raised up."

But there are other heavenly cycles which can also be seen as providing foundation for the myth of Osiris cast down and then raised back up. Another, as already discussed, is the annual cycle, with its interplay of light and dark, and its "two halves" to the year -- one in which hours of daylight dominate over hours of darkness, and one in which the opposite condition prevails. The two "crossing points" between these two halves of the year are the equinoxes: at the fall equinox, we reach the transition between the "upper half" of the year (in which daylight is longer than night-time) and the "lower half" (in which night-time prevails over daylight).

Thus, at the fall equinox, it can be said that the sun-god is "slain" or "conquered" -- cast down into the underworld. Of course, this slaying or this conquest can be seen as being *only temporary* -- because as the earth in its orbit arrives at the other equinox (in the spring), it will reach another point of "crossing over," after which the hours of daylight will begin to be longer than the hours of night-time once again, a condition that will last through spring and summer and the first part of fall.

As we have seen in two different quotations already, Alvin Boyd Kuhn argued that the point of autumnal equinox can be seen as the point of the "casting down" of the Djed -- the station of the year which marks the death of Osiris, and his banishment to the underworld.

According to this theory (and Kuhn backs up his arguments for this interpretation with extensive evidence from ancient myth), the beginning of the "restoration of the Djed" takes place at the point of winter solstice, when the "downward plunge" of the sun's path finally turns back around (and days stop growing shorter and begin to grow longer again). Thus the cessation of this downward motion marks the beginning of the restoration, but the full triumphant restoration is not achieved until spring equinox arrives, and the sun again "crosses back upwards" into the "upper half" of the year, when daylight hours are finally longer than hours of darkness.

Under this understanding, the death and resurrection of Osiris can be seen as reflecting the annual cycle of the orbit of earth around the sun, and the seasonal changes caused by the tilt of our planet's axis of rotation relative to our orbital plane. Thinking back to the zodiac diagrams presented earlier, the casting down of the sun-god takes and his path through the "underworld" begins at the point of autumn equinox, and lasts throughout the entire lower half of the circle. This understanding harmonizes well with the daily cycle, which the sun enacts each day by disappearing at sundown and then "plowing through the underworld" in order to reach the point of sunrise (which equates to spring equinox on the annual cycle).

And there is yet another way in which the annual cycle may be seen reflected in the Osiris myth, which is a function of the fact that Orion can be said to disappear entirely from the night sky for a period of about seventy days (the exact period will depend on the observer's latitude). It is also possible to argue that the "death of Osiris" and his eventual triumph over that death represents the disappearance of Orion from the night sky during this seventy-day period, and the constellation's eventual return.

This period of disappearance is also a function of the annual cycle, and is caused by the fact that the earth is constantly making progress along its orbital path, which causes the position of the stars to be slightly different on each successive night than they were the night before (because the platform upon which the observer is standing has moved a little bit through space).

This "forward progress" of the earth along its orbital path causes the constellations to be positioned a little bit further west on each successive night. That is to say: the stars not only move from east to west during the night (if you were to sit outside for several hours and observe them), a phenomena which is caused by the rotation of our globe upon its axis each day, but they *also* can be seen to make progress from east to west as we move throughout the year, a phenomena which is caused by the progress of our planet along its orbital track. Said in a different way in order to make this concept perfectly clear: the stars do indeed move from east to west each night as our planet rotates, but they actually "start" this westerly motion each night from a point that is slightly further west from their "start point" the night before. If the stars of the constellation Leo were just clearing the eastern horizon at 10 pm on January

20th, for example, they would be seen to be just a little further above the eastern horizon on the 21st at the exact same time of 10 pm, and a bit further along at 10 pm the succeeding night. In fact, the stars will be about four degrees of arc further west at the same time on each successive night, as a function of earth's progress along its orbit.

If you make a habit of going out to observe the stars every night or every morning (before sunup) at about the same time every day, you will observe this phenomena for yourself. As we move through the year, this motion will cause a constellation seen at the eastern horizon at 10 pm on one night to be a bit "further along" towards the west each succeeding night at the same time, until the constellation that was once rising up at the eastern horizon at 10 pm will actually be seen to be sinking down into the western horizon at 10 pm several weeks later.

If you think about it, you will understand why this "western progress" will take constellations out of sight for a certain portion of the year, as their setting time keeps moving forward each night until it actually occurs before the sun goes down: by sundown, the constellation itself has already set in the west.

The illustration below will illustrate the cause of this phenomenon, and will show us why Orion (and other constellations other than the circumpolar stars) will be out of view for a certain part of each year.

As you look at the diagram, remember that we can only see the stars in the night sky when we are on the side of the globe which is turned *away* from the sun (the night-time half of the planet). Thus, it should be fairly obvious that when earth is located on the portion of its orbit that is nearest to the viewer as we look at the diagram, the stars of Orion will actually be in the sky only when we are turned towards the sun -- that is to say, Orion will be in the sky during the daytime (and thus invisible).

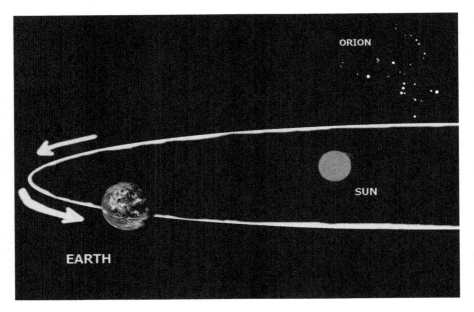

When the earth is at a location on the orbit that is on the "other side" of the sun (as we look at the diagram), then the stars of Orion will of course be visible during the night-time hours. On that side of the orbital track, observers on our planet will turn towards the location of Orion's stars in the sky as their part of the planet rotates around to face away from the sun (night-time).

But note the direction of the earth's progress along its orbital path, indicated by the arrows at the left side of the diagram. Each night, as earth moves along the path in this direction, the stars of Orion will get just a little bit "further back" as earth rotates on its axis towards the east -- which means that the constellation itself will appear just a bit further towards the west each night. Eventually, as earth turns the corner on the diagram, Orion will be seen to be setting in the west very soon after the sun itself sets in the west.

When the earth reaches the place on the orbital track where it is depicted in the diagram, Orion will begin to be out of sight: it will be setting almost concurrently with the sun. All along the path we see "to the right" of the earth at its location in the diagram, Orion will not be visible from earth: the sun will be "in the way." However, as earth continues its progress, its daily turning will again reveal the stars of Orion -- and if you look at the diagram and imagine the earth spinning on its axis and continuing along its orbital path, you may be able to envision in your mind the situation in which Orion will begin to be visible again for the first time after this period of "invisibility."

(Of course, it helps if you know how to envision the earth spinning in the proper direction; in order to envision the rotation properly, think of the continent of North America, and envision it on the globe, with north being "up" in our diagram: the east coast of North America will "see the sun" first each morning, and will have a time zone that is three hours ahead of the west coast, because the globe itself is spinning towards the east). This will help you envision the direction the earth spins.

If you are using your mind's eye, you will be able to envision the fact that as earth gets further and further along its track, its forward progress will eventually bring it to the point at which its eastward spin will begin to reveal the stars of Orion again, just *before* the eastward spin brings the sun into view. This phenomenon is known as "heliacal rising," when a star's daily progress towards the west finally brings it above the horizon just prior to sunrise. Each day, the star will have a tiny "head start" over its position the previous day at the same time, such that the star that was just above the eastern horizon at sunrise on one day will be a tiny bit further above it (further west) at the same time the next day, and so on through the year. Thus, Orion's entire form will begin to rise a bit earlier each night, until he is no longer rising just prior to sunrise, but will rise closer and closer to midnight as the months progress, until he is rising in the hours before midnight (during the winter months). At this point, the earth will be back around to the "other side" of the orbital track, in our little diagram on the earlier page.

Thus, we see yet another way in which the celestial mechanics can be said to cause Orion (Osiris) to "die" in the west, concurrent with the sunset, only to be "born again" in the east, concurrent with the sunrise. The earth's daily motion (spinning on its axis) as well as its annual motion (progressing along its orbital path) can both be seen to act in harmony with one another, causing Orion (Osiris) to "plunge down" into a sort of "underworld" or "realm of the dead," only to re-emerge later on, after a predictable interval.

And yet there are still more heavenly cycles which can be seen to be related to the myth of the dying god and his return -- and one of these is by far the most important with respect to the Osiris myth and with respect to other Star Myths from around the world which relate to message of the Osiris myth and which use the constellation Orion in the same way that the Egyptian myths use Orion.

That heavenly cycle is the ages-long motion known as *precession* (frequently referred to as the "precession of the equinoxes") -- the slowest of all the heavenly cycles used in myth (to my knowledge, at this point), and thus the most awe-inspiring and majestic of them all.

If we imagine the heavenly cycles as being turned by great celestial gears, then it stands to reason that the faster cycles will be turned by smaller gears, and the slower cycles will be turned by successively larger gears. Thus, the ages-long grinding of the precession of the equinoxes reveals to us the turning of the very largest of all the gears that can be observed with naked-eye astronomy.

In fact, the motion of the gears of precession are so slow (and the gears themselves so seemingly vast) that precession is barely perceptible, even for a keen observer making precise observations and careful recordings for an entire human lifetime.

As mentioned previously, the cause of the phenomenon known as precession is generally ascribed to the "wobble" of the axis of earth's rotation -- the change in the direction that the axis of rotation (the line running through our planet's north and south poles) "points" towards in outer space. Walter Cruttenden has made observations and published arguments which may point to a different cause of precession -- but leaving the actual cause of the phenomenon aside for the purpose of this volume, we can describe the effect of this phenomenon upon the stars of the night sky. Precession's impact upon the night sky is fairly easy to describe:

The motion of precession causes a delay in the entire "background of stars" from one year to the next.

This delay is best understood in light of the foregoing discussion regarding the earth's progress around its orbital track, in which we noted that the same star or group of stars will be slightly further along on each successive night, due to the earth's continued progress along its orbit.

If you observe the stars each night for a year, you will observe this westward progress for yourself -- and if you do so very precisely and very carefully, perhaps over the course of several years, you can actually build up a record of where a star will be in the sky at a precise time and day throughout the year.

In other words, you should be able to predict the day when the first star of Orion will be visible above the eastern horizon at a specific time of the morning before sunrise. You should also be able to predict the exact day and time in which a particular star in Orion will be crossing the very highest point of its arc on any given day of the year.

Because the earth goes around its track each year, you should see the same stars in exactly the same positions if you are standing in the exact same spot on the planet at the exact same point in time that you were standing when the earth got to the exact same point in its orbit the prior year.

And, for the most part, you will.

But because precession acts to delay the background of stars over a very long period of time, after a long enough period of years, you will find that the stars on any particular date and time in the year will be just a little bit behind where you observed them *at that same particular date and time* several decades prior.

In other words, if a specific star in the constellation Orion is exactly 12 degrees above the eastern horizon at 4:00 am on a day when the earth has reached its exact same point on the orbital path that it reached one year prior, we might expect that the same star would always be exactly 12 degrees above the eastern horizon at 4:00 am on that same day. And, even if we observe the stars with great precision for many years in a row, it will appear that they *do* follow this predictable pattern, returning to their exact same spot at the appointed time on the same date each year.

However, the motion of precession, which very gradually *delays* the background of stars, will eventually upset this predictable return -- but it will take a very long time before a normal observer can detect this delay (without extremely precise instrumentation, optics, and measuring tools).

Note that our *calendar* actually "slips around" a little bit from year to year, due to the fact that earth doesn't make a perfectly even number of turns as it gets back to the same point on its orbit -- which is why the solstices and equinoxes are such important "milestones" on our orbital path, and why we insert an extra day during "leap years" in order to keep the calendar from slipping too far off. However, even if we use the day of spring equinox *itself,* for example, ensuring that we are back to the *exact* same point in our orbit where we were at the previous spring equinox, the motion of precession (independent of any calendar slippage) will cause the stars to be just a little bit behind their expected place on that particular day -- but only by a single degree after 71.6 years.

This "precessional delay" will not really make much difference, even after 72 years: if the stars of Orion were above the eastern horizon at the sunrise of spring equinox seventy-two years ago, they will still be in almost the exact same position on the sunrise of spring equinox this year, except that they will all be just one degree lower (further east) than they were on the same day seventy-two years before.

One degree is barely noticeable -- especially if it takes seventy-two years for that degree of delay to show up. However, after thousands of years, that slow delaying motion will mean that the stars of Orion which once were above the horizon on the morning of spring equinox will be so far delayed that they will still be below the horizon on the day of spring equinox -- as if they are "over-sleeping."

Thus, throughout the course of millennia, Orion's date of heliacal return will be later and later in the year. Thousands of years ago, during the Age of Gemini, Orion was making his annual "reappearance" after his annual period of "invisibility" at the time of the spring equinox. But due to the inexorable motion of precession, the stars of Orion were eventually delayed to the point that the constellation did *not* rise above the horizon by the date of spring equinox.

Over the centuries, Orion's heliacal return came a bit later and later during the year. Now, thousands of years later, Orion and Gemini are above the eastern horizon just prior to sunrise during the beginning of the month of August -- much later than the spring equinox in the northern hemisphere, which falls in the month of March.

As has been capably argued by Giorgio de Santillana and Hertha von Dechend in *Hamlet's Mill*, and by Jane B. Sellers in *Death of Gods in Ancient Egypt*, this figurative "holding down" of the stars of Orion below the eastern horizon on the expected day of spring equinox can be metaphorically embodied in myth by saying that Osiris is being *drowned*, *buried*, or otherwise *murdered* by conspiratorial forces – and this is exactly what the ancient myths describe as being done to the god.

In fact, as Jane B. Sellers points out, the Isis and Osiris myth as recorded by Plutarch around the end of the first century AD describes the murder of Osiris (and his being cast into the sea) by his brother Set or Seth and *seventy-two* henchmen -- an astonishingly accurate precessional number! And, while Plutarch is indeed a later source, it is quite possible that the tradition he is relating originated long before -- because such an accurate estimation of the precessional constant was not known by *any* of the astronomers of Plutarch's day, as far as we know from the historical record.[34]

In fact, the Osiris story is almost certainly designed to encode the precessional cycle as well as the much shorter annual and daily cycles that we have been examining. In addition to the detail related by Plutarch regarding the number of henchmen who assist Set in the treacherous murder of the god Osiris, there is also the fact of his being cast into the sea, as well as the fact of his association with the constellation Orion, which was "held down" beneath the horizon on the morning of spring equinox during the time that the Age of Gemini was giving way to the Age of *Taurus* -- and, as Jane B. Sellers also points out, there are some reasons to believe that Taurus and the red star of Aldebaran were anciently associated with the malevolent god Set, slayer of Osiris. Also, during the judgment of the Ennead, Set was punished by being placed under Osiris, perhaps in the form of the constellation Lepus, just below Orion, whose outline does indeed resemble the shape of the "Set-beast."[35]

Further, as I have argued in previous books based partly on the arguments presented in *Hamlet's Mill*, elements of the Osiris story such as the "chopping down" of the central pillar (in which a tree grows around the casket of Osiris, and is then used as a pillar in the palace of the king and queen of Byblos) can be seen to embody themes of precession found in other precessional myths around the world.[36]

But perhaps the most powerful evidence which supports the conclusion that the myth of the death and restoration of Osiris is strongly precessional in nature comes from the fact that we can see a clear echo of the Orion association in other sacred texts from other cultures – sacred texts that can also be shown to possess explicit precessional references.

But these echoes can only be heard if we know our "astrotheology."

For example, as discussed in *Star Myths of the World, Volume Two*, the dislocation of Odysseus in the well-known story of the Odyssey, in which the hero languishes on the island of Ogygia for many years before he is finally permitted to make his way back home, can be shown to have clear precessional components. Perhaps the most telling of all of them is the fact that the number of the suitors who are plaguing his household in his absence and trying to woo his wife and also plotting to kill his son, a number which can be obtained based on a list given in the lines of the poem in Book Sixteen, totals to one hundred eight -- a precessional number that is one-and-a-half times the precessional constant of seventy-two, and which appears in myths and sacred traditions around the globe.[37]

Additionally, although my analysis in *Star Myths of the World, Volume Two* shows that Odysseus is most closely associated with the constellation Hercules, there are extremely important parts of the Odyssey in which he is unmistakably identified with Orion -- the same constellation with which Osiris is associated.

One such episode is the famous "trial of the bow" upon his return to his home, while he is still in disguise -- the very moment in which he casts off his disguise and reveals his true identity as the king who was lost and presumed dead but who has now returned. Based on my analysis, presented in *Star Myths Volume Two*, I show that this famous scene (which has given scholars so much trouble, as they try to figure out how anyone could shoot an arrow through the axe-heads set in the earth as described) is actually based upon a scene *in the heavens*, which solves the dilemma that has puzzled previous interpreters of the bow-trial's layout.[38] In that celestial scene, as I interpret it in that volume, Odysseus is clearly associated with the figure of Orion, a constellation which indeed does appear to be holding a mighty bow in his outstretched forward arm.

An even more important association of Odysseus with the outline of Orion, however, is found in the instructions given to the long-suffering hero in Book Ten of the Odyssey, when he makes his way to the land of the dead to consult the shade of the great seer Tiresias. There, the shade of Tiresias tells Odysseus that he may eventually return home, although it will be after the destruction of his own ship and the death of all his companions, in a foreign ship, very late, and in calamity -- and that when he finally arrives home he will find a troubled house, filled with overbearing men, wooing his own goddess-like wife.

Then Tiresias tells Odysseus what to do at that point, giving him very distinctive instructions, containing strong parallels to other important celestial myths (the following translation is from the literal translation by Theodore Alois Buckley, published in 1896):

> But certainly when thou comest thou wilt revenge their violence, but when thou slayest the suitors in thy palace, either by deceit, or openly with sharp brass, then go, taking a well-fitted oar, until thou comest to those men, who are not acquainted with the sea, nor eat food mixed with salt, nor indeed are acquainted with crimson-cheeked ships, nor well-fitted oars, which also are wings to ships. But I will tell thee a very manifest sign, nor will it escape thee: when another traveller, now meeting thee, shall say that thou has a winnowing fan on thy illustrious shoulder, then at length having fixed thy well-fitted oar in the earth, and having offered beautiful sacrifices to King Neptune [Poseidon], a ram, and bull, and boar, the mate of swine, return home, and offer up sacred hecatombs to the immortal gods, who possess the wide heaven, to all in order; but death will come upon thee away from the sea, gentle, very much such a one, as will kill thee, taken with gentle old age; and the people around thee will be happy: these things I tell thee true.[39]

As I explain at some length in *Star Myths of the World, Volume Two* (which explores the celestial foundation of the sacred myths of ancient Greece, including many of the important events in the Iliad and the Odyssey), this famous injunction to Odysseus from the shade of the departed seer Tiresias is almost certainly celestial in origin: the oar which Odysseus is to carry upon his illustrious shoulder, which will eventually be mistaken for a "winnowing fan," is a direct reference to the constellation Orion -- and very much connects the theme of his long banishment and return with the story of

Osiris (who is also associated with the constellation, and who is also cast into the sea).

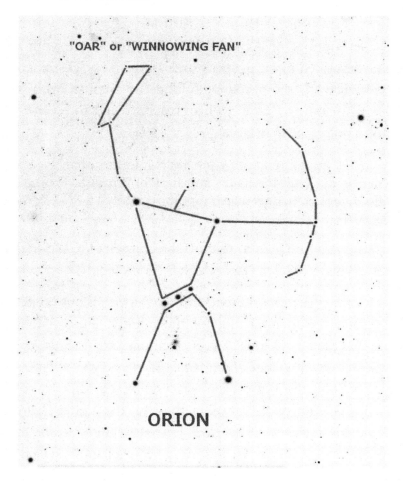

As you can see from the accompanying image, the outline of the constellation Orion features an "upraised arm" which terminates in an oblong rectangle formed by four faint but distinct stars. It is my contention that this upraised arm features in some Star Myths from around the world as the "oar" or "paddle" which certain important figures carry, and which surfaces in the ancient story of the Odyssey when Tiresias instructs Odysseus to take a well-fitted oar and carry it upon his shoulder until he comes to a land where the inhabitants know nothing of the sea -- and where they describe the oar as a "fan" with which to winnow grain.

This description, of course, has strong resonance with the events described in the gospel stories of the New Testament -- specifically with the account of John the Baptist, who tells those who come to listen to his preaching and to receive his baptism unto repentance that:

> "[. . .] he that cometh after me is mightier than I, [. . .] whose fan is in his hand, and he will throughly purge his floor, and gather his wheat into the garner [. . .]" (Matthew 3: 11 - 12).

This again can be seen to refer to the constellation Orion and to the upraised "oar" or "fan" which is "in his hand."

Indeed, this description found in the gospels all but confirms for certain the identification of the mighty figure with the winnowing fan as Orion -- for as I demonstrate in *Star Myths of the World, Volume Three* (*Star Myths of the Bible*), the figure of John the Baptist almost certainly corresponds with the constellation of Aquarius in the sky (the Water-Bearer), and the constellation Orion does indeed follow or *come after* Aquarius, and Orion is indeed a constellation which is "mightier" in appearance than Aquarius (particularly in terms of bright stars).[40]

It should not be difficult to see that the gospel accounts of the Christ contain strong parallels to the story of Osiris, the dying god who is "cast down" but who is later "raised back up."

It should not be too much more difficult to see that the story of Odysseus contains many of the very same parallels. In fact, just as a tree is central to the myth-cycle of Osiris, and to the story of the Christ (the cross itself being often referred to as "the tree"), so also is a tree and its threatened "chopping down" central to the climax of the story of the Odyssey, when Penelope tells Odysseus that she has moved their marriage bed -- eliciting his outrage and revealing his true identity to his long-suffering wife (she has not actually moved his bed at all: one of its posts is in fact a living tree-trunk, as Odysseus exclaims in his consternation).

All of these Star Myths, from very different cultures, clearly appear to share common celestial foundations – and clearly appear to be intended to convey a very similar message. They all deal with a difficult period of exile, characterized by suffering and betrayal and a passage through "the underworld" -- as well as with an eventual restoration and triumphant return.

As we have already seen from some of the quotations offered above from Alvin Boyd Kuhn, it is very likely that these ancient myths were not intended to describe literal and historical events from the distant past, events which happened to someone else who is qualitatively different from ourselves, but rather to convey to our understanding truths about *our own condition in this incarnate life*, right here and right now.

The story of the "cast-down Djed," the story of the "dying god," is the story of our own soul, our own divine spiritual element, cast down into this "underworld" of matter, where we are undergoing the same difficult exile.

But, as all the stories show, the Djed must be raised back up -- we must remember who we are, and elevate that divine aspect of our nature. The ancient myths and sacred stories indicate that this task is to be one of our central purposes during this arduous journey through the lower realm -- and what is more, they provide us with practical wisdom to help us in that essential task.

The Failed Baptism

In her collection of myths and sacred stories about the semi-divine culture-hero Maui, entitled *Maui of a Thousand Tricks: His Oceanic and European Biographers* and published in 1946, Katharine Luomala presents stories and oral traditions about the life of Maui collected from islands scattered across thousands of miles of the mighty Pacific Ocean, from Hawai'i to Aotearoa, from the Tongan and Fijian and Samoan Archipelagoes to Rotuma, the Tuamotus, Pukapuka, and Rakahanga.

One of the important episodes in the life of Maui, found in many versions of the Maui cycle, concerns the baptism of Maui by his father. In one of these versions, related by the Arawa tribe of Aotearoa (New Zealand), Maui and his brothers notice that their mother disappears into the ground each night -- so Maui makes note of the spot where she always goes, and then turns himself into a beautiful wood pigeon and flies down the long, narrow cave-tunnel that leads straight down into the underworld, where he eventually perches in a tree and sees his parents below him, his mother lying down beside his father on the ground beneath the tree, according to the story.

Maui, still in the form of a bird, then throws berries down at his father until his father throws a rock that knocks the bird out of the tree, and Maui turns back into his original shape. At this point in the story, Katharine Luomala tells us:

> Then the lad was taken by his father to the water to be baptized. And after the ceremony prayers were offered to make him sacred and clean from all impurities. But when it was completed, his father Makea-tu-tara felt greatly alarmed, because he remembered that he had,

105

from mistake, hurriedly skipped over part of the prayers of the baptismal service to purify Maui. He knew that the gods would be certain to punish this fault by causing Maui to die. His alarm and anxiety were therefore extreme. At nightfall they all went into his house.[41]

Some analysts have seen in this episode clear evidence of Christian influence upon the myths of the Maui cycle, since baptism is supposedly a Biblical concept that must have arrived from outside visitors -- but we should not be overly hasty in accepting this assertion as the only possible explanation for the evidence we find.

In fact, we find a story with remarkable similarities to this "failed baptism" (which results in Maui's mortality and susceptibility to eventual death) in the story of the semi-divine Achilles and *his* parents, who also attempt to "baptize" their infant son in order to confer invulnerability upon the child, only to have the process spoiled by a failure to dip his heel in the baptismal source (fire in some versions, and the River Styx in other versions).

Of course, the Achilles stories are not "Biblical" in their source -- and thus it is quite impossible to argue that every myth and story around the world which follows this pattern must necessarily have gotten it from Christian missionaries!

Another argument against the certainty of a later Christian influence upon the story of Maui's failed baptism is the fact (not widely known but demonstrable using the evidence in the stars themselves) that the entire "failed baptism" motif can be shown to reflect patterns in the night sky. Would Christian missionaries be likely to have suggested those aspects of the "failed baptism" episode to the originators of the Maui story?

We can agree that this possibility seems most unlikely (as does the suggestion that Christian missionaries suggested the outline of a story that parallels more closely the myth of the godlike hero Achilles than any story we find in the Old or New Testaments of the Bible).

The story of the "failed baptism" of Achilles is very well known and was clearly well known in antiquity. Although the baptism of Achilles is not a feature in the Iliad, the oldest textual source of our knowledge of the Achilles tradition, it was described or mentioned in passing by numerous other ancient authors, including Apollodorus, Lycophron, Statius, and many others.
In most accounts, Achilles is dipped into fire by his mother, the divine Thetis, in order to render him immortal – but is pulled out of the fire by his horrified mortal father, Peleus, thus leaving him mortal and eventually leading to his death in battle at Troy.

In other accounts, Achilles is dipped instead into the River Styx by his mother, also in order to render him mortal. In this version, Achilles' heel remains vulnerable, because when his mother dipped him into the dreaded waters of Styx, she held him by the heel, and thus that heel remained the one part of his body where an enemy could deliver a mortal wound. According to accounts preserved by some ancient authors, the god Apollo guided Paris the son of Priam of Troy to aim his arrow at this one point on Achilles' anatomy, and thus brought down the hero.

The similarity to the failed baptism of Maui should be self-evident. The similarity is especially pronounced in the versions of the Achilles myth in which the infant is dipped in the fire by his divine mother, but pulled out prematurely by his mortal father, because in the Maui myth it is the father whose oversight during the baptismal recitation leads to the mortality and eventual death of the semi-divine Maui.

It is also appropriate to bring up another story in the Maui cycle of myth, earlier in the life of Maui -- when he is just newly born. In many islands, there is a received oral tradition that Maui's parents threw him away when he had just been born (in some versions of the story, it is because Maui has eight heads). But he is rescued and resuscitated by his grandfather, who hangs the baby up to dry among the rafters of his home, above a great fire for warmth.

Here is the account as told among the Arawa of Aotearoa and recorded in Katharine Luomala's *Maui of a Thousand Tricks* -- Maui is speaking to his mother:

> "I knew I was born at the side of the sea, and was thrown by you into the foam of the surf, after you had wrapped me up in a tuft of your hair, which you cut off for the purpose. Then the sea-weed formed and fashioned me, as caught in its long tangles the ever-heaving surges of the sea rolled me, folded as I was in them, from side to side. At length the breezes and squalls which blew from the ocean drifted me on shore again, and the soft jelly-fish of the long sandy beaches rolled themselves round me to protect me. Then again myriads of flies alighted on me to buzz about me and lay their eggs, that maggots might eat me, and flocks of birds collected round me to peck me to pieces. But, at that moment, appeared there also my great ancestor Tama-nui-ki-te-Rangi. he saw the flies and the birds collected in clusters and flocks above the jelly-fish. Behold, within there lay a human being. Then he caught me up and carried me to his house, he hung me up in the roof that I might feel the warm smoke and the heat of the fire. Thus I was saved alive by the kindness of that old man."[42]

All of this snatching up of infants and suspending them in rivers and above fires can be seen to be related -- and related by reference to the same set of constellations in the heavens, according to my analysis.

It is not inherently obvious to the casual observer, but the glistening arc of the Northern Crown (Corona Borealis) plays the role of an infant in numerous Star Myths from around the world -- often an infant who is "snatched up" by a character corresponding to the powerful figure of the constellation Hercules, which is located very close to Corona Borealis in the sky, and facing towards it (such that it only takes a little imagination to envision a line connecting the forward-side arm of Hercules reaching out to grasp the Crown, as shown on the following page).

That the constellation Corona Borealis would not normally suggest itself as an infant is a very strong argument against the theory that all these various Star Myths in different cultures simply "sprang up" independently of one another -- because the Northern Crown shows up in so many different cultures in the form of a baby. We might expect it to show up as a crown, or as a necklace, in different myths independently, because its shape suggests those items without question (and it does show up in numerous myths as a necklace, for example). But the fact that it shows up multiple times as a dangling infant suggests some sort of ancient connection between the myths of these various cultures spread around the globe -- most likely, in my opinion, a common source of even greater antiquity, perhaps a common world-wide culture or civilization predating known history by thousands of years.

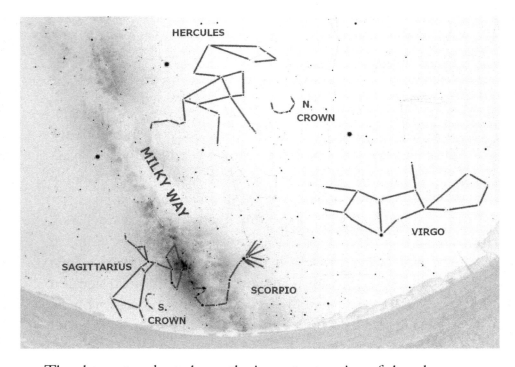

The above star-chart shows the important region of sky where the Milky Way rises between Scorpio and Sagittarius and proceeds up past Hercules. Constellation outlines have been added for those three constellations, as well as for Virgo and for the Northern and Southern Crowns. The outlines mainly follow those suggested by H. A. Rey -- but Scorpio has been shown in an alternate outline to illustrate the reason that this constellation appears in many ancient myths as a serpent with multiple heads (sometimes seven, sometimes eight, and sometimes nine).

Recall that in the Maui cycle of myths, Maui is thrown by his mother into the sea-foam as soon as he is born -- and then note the figure of Virgo to the right of the form of Scorpio as we face the page. The position of Virgo, and the fact that in the northern hemisphere she appears to be positioned on her back and with her legs elevated and spread apart, gives rise to the many myths in which she is envisioned as being about to give birth, or in the act of giving birth.

In some versions of the Maui myth, the infant Maui has eight heads. Katharine Luomala relates an account from the Society Islands (the archipelago in the South Pacific containing Tahiti, Moorea, Raiatea, Bora Bora, and many other islands) in which:

> Maui, prematurely born and enveloped in the placenta, looks like a jellyfish. His parents not knowing there is a living thing in the ugly birth, wrap it in a girdle of breadfruit bark, tie it with a hairdress (*ti'iti'i*), and cast it into the sea with a prayer.[43]

Because of this, he is known among the cultures of the Society Islands as "Maui-tiitii" and as "Maui-of-the-placenta and Maui-of-the-eight-heads."[44] Note that the word *ti'i-t'i'i* is a version of *tiki-tiki* (top-knot, hair-bun) in which the "k" sound has been elided -- just as I believe that the name of the island far to the east, Hawai'i, is a linguistically elided version of the name of the legendary origin-land of Hawaiki.

It should be obvious that the origin of the tradition of Maui's eight heads comes from the fact that the multi-headed form of the constellation Scorpio is positioned not far from and just below the upraised legs of the supine form of Virgo. Scorpio is also enveloped in the widest and brightest portion of the Milky Way galaxy -- which can be seen as resembling sea-foam (see for instance the analysis in *Star Myths of the World, Volume Two*, which presents evidence that the Milky Way plays the role of the sea-shore where the Achaean ships are pulled up onto the sand, and where the Achaean warriors are encamped during the battle of Troy). This is where Maui's parents threw him, when they cast the bundle "into the sea with a prayer."

The grandfather who pulls Maui out of the sea-foam and warms him above the fire can now be identified – he is an embodiment of the powerful figure of Hercules, high above Scorpio. Just in front of Hercules, we see the brilliant arc of stars that form the Northern Crown, Corona Borealis. This arc also represents the infant Maui, when his grandfather pulls him out of the ocean and hangs him up to dry above the fire.

Although the outlines of the figures as suggested by H. A. Rey do not include a "line" between the lower arm of Hercules (the arm not holding his sword or club) and the arc of the Northern Crown, we are certainly at liberty to envision such a connecting line – and the ancient mythographers clearly did so. There are numerous myths in which he grasps the Crown, as we will see.

The fact that Maui's grandfather hoists the infant up to the roof of his home is no doubt accounted for by the fact that both Hercules and Corona Borealis are very northern constellations, located near the celestial north pole. Of course, this also suggests the possibility that the origin-point of this ancient worldwide myth-system was located in the northern hemisphere – and as I've said before, I believe the origin stretches back to a time of great antiquity.

The fact that Maui's grandfather is described as hoisting the infant up *above the fire* in order to warm the little Maui is no doubt accounted for by the fact that the constellation Hercules is located just beside the rising column of the Milky Way itself, which does indeed resemble a rising column of smoke from a fire (and plays the role of smoke from a fire in countless Star Myths from around the globe).

We've now seen, in the Maui story, the shining column of the Milky Way playing two different roles -- first as sea-foam (water), and then as smoke (rising up from fire). Now, note the two versions of the Achilles tradition, in which his mother dips the baby Achilles in either a *fire*, or in the *River* Styx, in order to render him immortal. In either case, I would argue that we are again dealing with the same region of the sky, and that the river / fire is played by the Milky Way galaxy.

In the case of the Achilles story, his mother Thetis is most likely a goddess associated with the constellation Sagittarius, which often appears in mythology as a female figure (and just as often as a male figure). It is possible that the constellation of the Southern Crown plays the role of the infant Achilles when placed into the fire (before his father snatches him out) -- although once he is snatched out of the fire by his father (played by the constellation Hercules), then the baby Achilles will be represented by the same arc of stars found in the Northern Crown which played the role of Maui when snatched out of the ocean by his grandfather, Tama.

The association of the constellation Corona Borealis with an infant being held by the heel is confirmed by a very different story, found in the Old Testament of the Bible. There, in the story that has come to be known as the "Judgment of Solomon" (in the book of 1 Kings, chapter 3), we see two mothers bringing two babies (one baby being alive, and the other being dead) before King Solomon, each arguing that the living baby belongs to her.

In that dramatic confrontation, Solomon orders the living child to be divided "in twain," and one half of the baby to be given to each mother. From the text, we know that Solomon must be directing someone else to actually split the baby in twain, because when the actual mother of the living baby cries out to spare the child, Solomon commands the swordsman to stop, and to give the living child to her.

The swordsman whom Solomon is addressing in this famous episode is again the constellation Hercules, and the living baby that the swordsman holds as he awaits the command of the king is again the heavenly figure of Corona Borealis. Artists down through the ages have preserved the "arch" of the Crown and the posture of Hercules in their depictions of the scene:

above: depiction by artist Peter Paul Rubens (1577 - 1640).

below: depiction by artist James Jacques Joseph Tissot (1836 - 1902).

Under the conventional paradigm of human history, it would be quite remarkable to find myths separated by such vast distances and great gulfs of time as those of ancient Greece (Achilles) and those of the Pacific islands (Maui) sharing such undeniable parallels (not to mention the Old Testament account of King Solomon). Even the fact that these myths are based on the stars (which are, of course, visible in the Pacific just as they are in Greece) does not explain the amazing fact that the stars of the constellations Hercules and Corona Borealis show up in both cultures in the form of myths about a "failed baptism."

It is not always easy to find the figure of Hercules, unless you know the outlining system suggested by H. A. Rey -- to argue that two completely isolated cultures both saw these stars in the very same way, and that they then saw the arch of the Northern Crown as an infant, and that they then (completely independently of one another) turned those stars into stories about a "failed baptism" that robbed the infant of immortality is simply so improbable that it must rank very low on the list of possible explanations.

In light of all the other evidence showing that myths around the world appear to be based upon a common system of celestial metaphor, we must conclude that the myths of the failed baptisms of Maui and of Achilles (as well as the story of the Judgment of Solomon from the Old Testament) are built upon a system of "mythologizing" the stars that somehow operates worldwide.

There is yet another example of the "failed baptism" pattern in ancient myth, from the Isis and Osiris story which we encountered in the previous chapter. In the account of the Isis and Osiris cycle of myths recorded by the philosopher Plutarch (thought to have lived from about AD 45 or 46 to about AD 120), after Set treacherously seals Osiris in a casket and casts him out to see, the goddess Isis searches all over the world for the body of her beloved.

Finally, by divine inspiration, the goddess ascertained that the coffin containing the body of Osiris had come to rest in Byblos, where a tree grew up around it and was then cut down and turned into a pillar to support the roof of the palace of the king and queen of that land.

Isis then went to the country of Byblos and, without revealing her identity, went and sat down in dejection beside a spring, where the queen's maidservants came upon her, and the goddess braided their hair for them and imparted a wondrous perfume of divine ambrosia to their bodies, such that the queen herself longed to have "such hairdressing" and such fragrance. Plutarch continues:

> Thus it happened that Isis was sent for and became so intimate with the queen that the queen made her the nurse of her baby. They say that the king's name was Malcander; the queen's name some say was Astartĕ, others Saosis, and still others Nemanûs, which the Greeks would call Athenaïs.
>
> They relate that Isis nursed the child by giving it her finger to suck instead of her breast, and in the night she would burn away the mortal portions of its body. She herself would turn into a swallow and flit about the pillar with a wailing lament, until the queen who had been watching, when she saw her babe on fire, gave forth a loud cry and thus deprived it of its immortality. Then the goddess disclosed herself and asked for the pillar which served to support the roof. She removed it with the greatest ease and cut away the wood of the heather which surrounded the chest; then, when she had wrapped up the wood in a linen cloth and had poured perfume upon it, she entrusted it to the care of the kings; and even to this day the people of Byblus venerate this wood which is preserved in the shrine of Isis.[45]

Once again, we see very clear parallels to the other "failed baptism" myths which we have already examined. The baby is placed in the fire, but this time it is the mother of the baby whose terror at seeing the child on fire deprives the infant of potential immortality. The additional details in this story involving Isis flying around the pillar in the form of a swallow, as well as her wrapping up the wooden coffin in linen cloth anointed with perfume are also celestial in nature, according to my analysis:

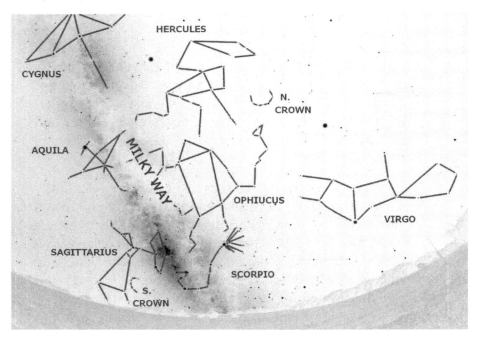

First, when Isis goes down to the spring in the land of Byblos (located in modern-day Lebanon, north of Beirut), she is undoubtedly connected in this part of the story with the figure of Sagittarius, located beside the widening part of the Milky Way band which appears as a spring or bathing pool in numerous ancient myths (many of them described in *Star Myths of the World, Volume Two*).

The maidservants of the queen coming down to the spring are probably identified with Scorpio's multiple heads -- this is a pattern we find again and again in the Greek myths, such as when Artemis goes to bathe along with her attendant nymphs in a secluded grotto, or when the princess Nausicaa goes with her attendants to a pool in a river in order to do the washing and is surprised by Odysseus (who has been sleeping in a bower between a great branching olive tree with two trunks).

Later, when Isis turns herself into a swallow and flits around the pillar made of the tree-trunk that has grown around her husband's casket, the myth is undoubtedly referencing one (or both) of the two great birds flying around the Milky Way galaxy just above Sagittarius and Scorpio: Aquila the Eagle and Cygnus the Swan (which have been outlined in the diagram on the preceding page).

Finally, when the king cuts down the pillar and the casket is removed from the tree-trunk, Plutarch tells us that according to ancient tradition, the goddess wrapped the wood in linen and anointed it with perfume. It is quite likely that the coffin-shaped outline of the constellation Ophiucus (also added to the star-chart on the preceding page) is playing the role of the coffin of Osiris in this part of the story, and that the two halves of what is usually envisioned as the "serpent" on either side of Ophiucus (which the constellation appears to be holding, hence its name of "the Serpent-bearer") are envisioned in this particular myth as the linen in which the goddess wraps the wood. Once again, there are other mythological precedents for seeing the "serpent" on either side of the central part of Ophiucus as linen or cloth (see for instance the story of the embarrassment of Draupadi in the Mahabharata of ancient India, discussed in *Star Myths of the World, Volume One*, as well as other myths discussed in that multi-volume series).[46]

There are also many myths in which I have identified the figure of Sagittarius as performing a role of "anointing" with oil, just as described in the account of the Isis and Osiris myth. Most of

119

this discussion can be found in *Star Myths of the World, Volume Three* (*Star Myths of the Bible*).[47]

In other words, the evidence supporting the identification of all these "failed baptism" myths with this specific region of the night sky is overwhelming.

But we may well ask ourselves, "What does it mean?"

I am convinced that these ancient stories have many layers of meaning, and that the message they have for our lives is profound and very pertinent to our situation, even in this day and age.

On one level, I am certain that these stories of "failed baptism" -- in which the struggle between mortality and immortality forms such a central theme -- all have to do with our descent into the material world, to take on an incarnate form, just as we saw in the previous chapter describing the "casting down" of the Djed column of Osiris.

Just as Osiris is the "dying god" (who will later be restored), we ourselves (the ancient myths seem to be telling us) are an admixture of mortal and immortal. We are possessed of an immortal nature -- a divine spark, an undying soul -- and yet we are plunged into a mortal body during our sojourn in this lower realm during this incarnate life.

Notice where the "action" of the myths discussed in this chapter takes place -- in the region of the sky between Scorpio and Sagittarius. The constellation Sagittarius marks the end of the "downward" journey of the sun, and the station just prior to the winter solstice, during the Age of Aries. In the precessional age immediately prior to the Age of Aries (during which much of ancient Egyptian civilization had its heyday), the winter solstice would have been between Scorpio and Sagittarius.

If we look back to the extended quotation from Alvin Boyd Kuhn cited on pages 55 to 57 of this book, you will see that the point of winter solstice was allegorized within the ancient system as marking a great "turning point," when the downward plunge into matter turned around, and an awakening to the reality of the invisible, spiritual realm is quickened in the heart. This point represents a sort of "new birth" or "second birth" according to Kuhn's interpretation of the ancient system (an interpretation which I believe to be quite correct).

Thus, this point on the great circle of the year represents a point of struggle between the divine nature of spirit and the mortal nature of the material form. We see this reflected in the myths of the baptism episode across the different cultures, from Egypt to Greece to the islands of the Pacific. Sometimes the child itself is semi-divine (Achilles is the son of a divine mother and a mortal father).

In each case, mortality is not completely "burned away" -- in this mortal existence, we each must struggle with the exigencies of life in the material realm, weighed down by a physical body and by our "lower" or "baser" passions and weaknesses, subject to the laws of physics, the "thousand natural shocks that Flesh is heir to" (as young prince Hamlet says in Act III, scene I of *Hamlet*).[48]

And yet, as the struggle between the mortal and the immortal depicted in the baptism myth shows, we are in some way supposed to be elevating the divine spark, in ourselves and in others -- "burning away" the lower distractions and obstacles to the elevating of the spiritual nature (the Yoga Sutras of

Patañjali, for example, identify obstacles and distractions such as nescience, self-importance, attachment, sensuality, laziness, illusion, and doubt, among others, as necessary to be overcome in this life, and urge the practice of *ahimsā* [not doing violence], *aparigraha* [not hoarding or collecting things beyond what is needed], and *asteya* [not desiring what another has], among other virtues).[49]

The passage up the Milky Way, or up the line running from the lowest point on the circle of the year (winter solstice) to the highest point on the same circle (summer solstice) is also associated with the "raised-up" Djed column (just as the line between the two equinox points is associated with the "cast down" or horizontal Djed, as shown on page 75).

Note that in the story of baby Maui, in which the grandfather Tama rescues the infant who has been abandoned in the sea, the direction of the story takes Maui upwards along this very line of elevation, from the castaway in Scorpio (with its multiple heads) straight upwards to Hercules, who hangs Maui up in the rafters (Corona Borealis).

This elevation "straight up the Djed column," so to speak, is an elevation along the "backbone of Osiris." It is not at all inappropriate to see that the incredible ancient wisdom also operates on an "individual" level: many ancient disciplines (such as Yoga and Yogic meditation, as well as Daoist meditation and other related disciplines) include practices designed to elevate the life-energy upwards along the spine, from the root chakra to the crown chakra (or similar parallels in traditions that do not use this exact terminology).

This is a theme to which we will return later in the book. Suffice it to say at this point that the elevation of the spiritual nature appears to be a central focus of many of the ancient traditions which have survived down through the millennia in some cultures in the form of disciplines such as Yoga, Tantra, the internal martial arts found in China and other parts of Asia, and other similar traditions.

Readers familiar with the scriptures of the New Testament may also have noticed during the preceding discussion that once we understand that the constellation Virgo was sometimes envisioned in ancient myth as a woman in labor, about to give birth or already in the act of giving birth, and that once we understand that the "multi-headed" form of Scorpio is positioned just beyond and below her upraised legs, then this understanding can help to shed light on the often seemingly-bizarre imagery of Apocalyptic literature such as the scenes described in the book of Revelation.

As mentioned in a previous chapter, in Revelation chapter 12, the vision describes "a great wonder in heaven" in the form of a woman who "being with child, travailing in birth" greatly desires to be delivered of the baby. Then "another wonder in heaven" appears, described as "a great red dragon, having seven heads and ten horns, and seven crowns upon his heads," and standing before the woman who is in the pains of labor, ready "to devour her child as soon as it was born."

Clearly, this scene describes the multi-headed form of Scorpio, stationed before and beneath the constellation Virgo, envisioned as a woman in heaven "travailing in birth."

When the woman does bring forth a child, however, instead of the child being devoured by the seven-headed dragon, the text tells us that "her child was caught up unto God, and to his throne."

This "catching up" or "snatching up" of the child is another manifestation of the same constellations we have been looking at: Corona Borealis caught up by the mighty figure of Hercules. This catching up of the child is sometimes interpreted by literal interpreters of the New Testament scriptures as being indicative of the "rapture" of the body of believers (the word "rapture" itself meaning to "snatch-up," and using the same root as is found in the word "raptor," used to describe birds of prey who snatch their prey in their powerful talons).

Rather than referring to literal future events (events in literal history, even if "future history"), I believe that the descriptions in the book of Revelation (along with the episodes described in the rest of the texts collected into what we today call "the Bible") are based on the same system of celestial metaphor, and are indeed intended to convey profound truths to our understanding – but that their message is metaphorical and not literal, and does not correspond to historical events (whether in the past or in the future).

In fact, I believe this "rapture" scene is referencing many of the very same themes conveyed to our understanding in the "failed baptism" episodes from around the world – and that it urges the same "elevation of the spirit" taught in other myths and ancient traditions -- the same goal towards the attainment of which disciplines such as Yoga and Tantra and meditation were developed and passed down from one generation to the next, for thousands of years.

Cast into the Sea: Crossing the Great Flood

In the previous chapter, we saw that Maui suffered a "failed baptism," but also that prior to this episode, baby Maui was mistaken for a jellyfish and cast into the sea by his parents (in some versions of the story, the fact that he has eight heads is also a possible reason for his rejection by his parents).

This aspect of the Maui myth resonates sympathetically with other myths found the world over, involving infants cast upon the waves for one reason or another.

One of the strongest echoes of the Maui myth is found in the oldest record of ancient Japan, known as the Kojiki (or Ko-Ji-Ki -- the syllables themselves correspond to the words:

古事記

which literally signify "ancient-matters-record," or "a record of ancient matters").

There, related without much elaborating detail, we find the story of the first child of the *Kami* (a word that means "deities" or "superior powers") Izanagi and Izanami. These two Kami, Izanagi being male and Izanami being female, immediately decide that they want to have sexual relations with one another, and agree to walk in a circular path around a pillar and in doing so to come together sexually when they meet. The text relates:

> Then Izanami-nö-mikötö said:
> "Then let us, you and me, walk in a circle around this heavenly pillar and meet and have conjugal intercourse."
> After thus agreeing, [Izanagi-nö-mikötö] then said:
> "You walk around from the right, and I will walk around from the left and meet you."

After having agreed to this, they circled around; then Izanami-nö-mikötö said first:

"*Ana-ni-yasi*, how good a lad!"

Afterwards, Izanagi-nö-mikötö said:

"*Ana-ni-yasi*, how good a maiden!"

After each had finished speaking, [Izanagi-nö-mikötö] said to his spouse:

"It is not proper that the woman speak first."

Nevertheless, they commenced procreation and gave birth to a leech-child. They placed this child into a boat made of reeds and floated it away.[50]

They later learn, through consultation with the higher heavenly powers, who use a form of scapulamancy (writing inquiries upon the dried shoulder-bone of an animal, in this case a deer, and then heating it to see where the cracks formed), that the reason that this first child was born without bones was in fact due to the impropriety of the woman's initiating the greetings first (in a strange echo to the story of Adam and Eve, in which the woman initiates the taking of the fruit -- a story that can also be shown to be based upon the constellations, including the initiative taken by Eve).[51]

The important point for the discussion at hand is the fact that this leech-child (whose name is actually

蛭子

which literally means "leech-child" and which is pronounced *Hiruko*, or *Hiru-Ko*, in Japanese) is cast into the waters in very much the same way that the baby Maui (born prematurely and resembling a jellyfish when wrapped in the placenta) is cast into the waters by his parents in the Polynesian myth.

In the Japanese mythology surrounding Hiruko, he was born without bones, and was thus unable to develop as other babies would, which is why his parents decided to construct "a boat made of reeds" and float him away.

Where else in mythology do we see an infant placed into a "boat made of reeds" and floated upon the waters?

Perhaps a better question to ask might be, "Where *don't* we see it?"

Of course, the most familiar manifestation of this story to western readers will almost certainly be the story of baby Moses, who is similarly floated upon the Nile River in an "ark made of bulrushes," as we read in Exodus chapter 2.

There, we are told that a man in the house of Levi (in captivity amongst the Egyptians) took to wife a daughter of Levi, and bare a son -- and after three months, when she could hide him no longer, "she took for him an ark of bulrushes, and daubed it with slime and with pitch, and put the child therein; and she laid *it* in the flags by the river's brink" (Exodus 2: 3).

It is perhaps noteworthy that in Japanese tradition, not specified in the Kojiki, it is after three years (when the boneless child still cannot walk) that Hiruko's parents place the child in the reed boat.

But Hiruko and Moses are by no means the only babes placed into a reed boat and cast adrift upon the waters. We find a very similar account in the ancient Sanskrit epics of India, where the baby Karna is similarly set adrift amongst the reeds by his mother Kunti in the Mahabharata. And, in some of the legends which grew up around the mighty figure of Sargon of Akkad (considered by some to have been an historical king), there is a birth-legend of Sargon considered by scholars to date to around 2300 BC, in which the king says:

> My mother, a priestess, conceived me and bore me in secret. She put me in a basket of reeds, sealed its lid with pitch; she cast me adrift on the river from which I could not arise. The river bore me up and brought me to Aqqi, a drawer of water.[52]

Those familiar with the scriptures of the Old Testament will note this consistent "daubing with pitch" or "sealing with pitch" found in the Moses and Sargon accounts, and will wonder if there may be a relation to the description of Noah, sealing the ark with pitch prior to the Genesis Flood. Indeed, there may *be* a connection.

Below we see the same region of the night sky we have been examining in the previous chapter. In the account from the Kojiki in which Izanagi and Izanami walk in a circle around a central pillar, I believe that it is very likely that the central pillar in the story represents the towering column of the Milky Way galaxy, rising between Sagittarius and Scorpio.

We are already fairly certain that Hiruko, the leech-child, is associated with the sinuous form of the constellation Scorpio, which does indeed seem to have "no bones" and to thus be unable to stand upright.

Note that the text of the Kojiki expressly specifies that Izanami will walk around the pillar "from the right" and Izanagi will walk around it "from the left." These instructions may indicate that the maiden Izanami is in this portion of the myth associated with the outline of Virgo the Virgin, which is indeed located to the right of the Milky Way as we face it in the sky (in the northern hemisphere):

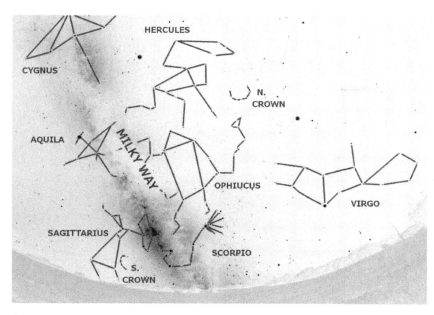

If that is the case, and if the instructions "from the right" and "from the left" indicate positions in relation to the central pillar of the Milky Way as we face it, then it is very likely that according to these directions, the male Izanagi corresponds to either the constellation Sagittarius (located to the left of the Milky Way as we face it in the northern hemisphere) or Aquarius (further to the left of Sagittarius, and about the same distance on the left side of the Milky Way that Virgo is distant to the right).

Note that this identification of Izanami with Virgo would comport well with the fact that Virgo is in the proper position to be seen as giving birth to Scorpio, the figure of which almost certainly corresponds to Hiruko. And the tentative identification of Izanagi (at least in this part of the story) with Aquarius would comport well with the fact that the "forward leg" of Aquarius can be shown to have been sometimes associated with a specific part of the male anatomy, which

129

Izanagi and Izanami discuss quite explicitly in the lines of the Kojiki, just before they agree to walk around the pillar in order to come together and have sexual relations.

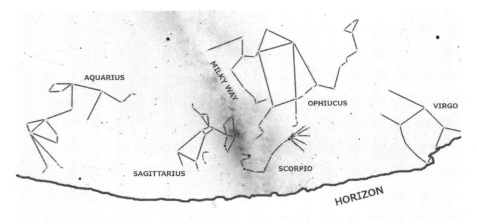

The identity of the "reed boat" itself, in which Hiruko is set adrift, has been argued in *Star Myths of the World, Volume One* to also possibly correspond to the constellation Scorpio, which does indeed resemble a "Viking ship"-type boat-outline (and which can be shown to play the role of a ship in other Star Myths from around the globe). It would not be unusual for Scorpio to play the roles both of Hiruko and then later of the boat in which Hiruko is set adrift (constellations often play more than one figure or character within the same myth).

It is also possible, and indeed even likely, that the "boat made of reeds" or the "basket among the reeds" in this and related "baby-set-adrift" stories correspond to the outline of the constellation Ophiucus in the sky. Note, for example, that in the text quoted above regarding the setting adrift of baby Moses, the infant's mother is described as placing her baby in an *ark* made of bulrushes. The shape of Ophiucus, with its boxy rectangular outline and triangular "roof" does indeed resemble the traditional shapes ascribed to "arks" in the Bible.

I believe that it is very likely that the "reeds" or "bulrushes" which are typically present in the stories about babies being set adrift may come from the two "serpent halves" on either side of the constellation Ophiucus -- these twisty lines often feature in myth as different types of vegetation: sometimes taking on the form of climbing grapevines, or of creepers, or of olive trunks, and so forth.

Additionally, the feature of some of the stories in which the mother daubs or anoints the ark with pitch or slime may well come from the figure of Sagittarius, whose "bow" (pointing into the Milky Way and hence towards Ophiucus) can be shown to sometimes play the role of a cruse of oil for anointing. Thus, the mother in the stories, even if generally associated with Virgo, may be envisioned as corresponding to Sagittarius when she actually daubs the ark with pitch and slime and then sets her child adrift.

But, the reader may well ask, "Why?" and "What does it mean?"

Indeed, why *would* these various myths from around the globe feature a mother casting away her precious child by placing the babe in a basket (or "ark") of reeds and setting it adrift upon the waters? What message could these stories possibly be intended to convey?

After much consideration, I have concluded that the message of these myths is related to the themes we have seen manifested in the previous patterns we've been examining -- as well as to the elevation of the spirit and the re-integration with the

Infinite Realm which we are supposed to be practicing during our sojourn in this incarnate life in the material realm.

We have already seen that there is a clear connection between the placing of Moses upon the water in an "ark of bulrushes" and the crossing of Noah and his family through the Genesis flood in another ark, this one pitched "within and without with pitch" (Genesis 6:14) – just as the ark in which the mother of Sargon of Akkad placed her baby was described as being sealed with pitch.

The clearest evidence of a thematic connection between the vessel in which Moses was cast adrift and that in which Noah and his family were cast adrift is the fact that in the scriptures themselves, the very same Hebrew word is used to describe both vessels:

תֵּבָה

a word of unknown etymology, Strong's number H8392, rendered *têbâh* in Latin letters and translated as "ark" (meaning a "chest or coffer") in the most-familiar English translations of the Old Testament scriptures.

The pitching of this chest with pitch, in both the accounts of Noah's ark and of the basket with a lid in which Sargon's mother placed her child, provides another clear connection between these stories.

Just as the Mesopotamian legends involving Sargon of Akkad find parallels in the story of Moses among the bulrushes in the Hebrew Scriptures, so also do we find ancient Mesopotamian texts describing flood narratives which find echoes in the account of the Genesis flood involving Noah. What's more, these Mesopotamian flood accounts provide us with important insights into the meaning and message of the flood-narrative pattern found in myths around the world, as well as the meaning and message of the "baby cast adrift" pattern which we have just seen to be closely related to the flood-myth pattern.

These accounts, including the account of Uta-napishti (or Utnapishtim) found in Tablet XI of the Gilgamesh series, and the related accounts of Ziusudra in a Sumerian version, and of Atrahasis in an Akkadian version, all contain very close parallels to the account of the Biblical deluge found in Genesis, including common elements such as the reason for the flood itself, the advanced warning given to a virtuous man who is instructed to build a box-like ark in which to ride out the flood with his family and with animals, the unleashing of waters for a certain number of days and nights (seven days and seven nights in most of the Mesopotamian accounts, and forty days and forty nights in the Genesis account), the opening of a window in the ark in order to release certain birds including a dove (to see if the waters have subsided enough to venture back out onto land), and the offering of thanksgiving once the flood has ended.

There is also, in both the Genesis account and the ancient Mesopotamian accounts, an explanation that the lifespan of men and women will be limited after the flood -- and the Mesopotamian accounts actually make the reason for this element of the flood story a little more understandable, tying it back to the reason the gods released the flood in the first place (because men and women were becoming too numerous and boisterous upon the earth, and keeping the gods from getting any rest).

In *Star Myths of the World, Volume Three* (Star Myths of the Bible), I provide some arguments that the ark of Noah described in the Bible may correspond to the "ship-like" shape of Scorpio, as well as to the "boxy" shape of the Great Square of Pegasus (especially when the ark comes to rest on Mount Ararat), and even perhaps to the form of Argo Navis (located as it is in the sky adjacent to Columba, the Dove).

Based on the fact that the very same word is used in the Hebrew text to describe both the ark of bulrushes in which Moses is cast adrift and the ark of Noah in which he and his family ride out the flood, I believe there is also evidence to argue that the oblong, rectangular form of Ophiucus may be a strong candidate to play the ark, at least during some parts of the flood-account. We will see from some other flood-myths found in other parts of the world that Ophiucus is definitely associated with the deluge -- and with the form of a basket (connecting again the flood and the stories of the child cast adrift).

Citing two volumes from the nineteenth century which record some of the myths of the indigenous tribes of what is today Guyana in the continent of South America (specifically, the Akawaio), the authors of *Hamlet's Mill* relate the story of the benevolent deity Sigu, son of the great spirit Makunaima, who is engaged in the task of spreading good seeds to all the lands of the earth, and spreading fish to all the waters of the earth, when a rising tide of water from a hollow tree stump threatens to overflow and cover the lands -- and so Sigu covers the stump with a "closely woven basket" to keep the waters from coming out.

Citing their nineteenth century sources, the authors of *Hamlet's Mill* explain:

> But unfortunately the brown monkey [the one animal in all of creation who refused to help Sigu, and who was therefore given the task of retrieving water from a stream in a leaky basket], tired of his fruitless task, stealthily returned, and his curiosity being aroused by the sight of the basket turned upside down, he imagined that it must conceal something good to eat. So he cautiously lifted it and peeped beneath, and *out poured the flood*, sweeping the monkey himself away and inundating the whole land.[53]

Note that here again we have the presence of a "basket" -- a container or a vessel made by plaiting together reeds or withes, very reminiscent of the basket in which the mothers of Karna and Sargon place their babies, or the "ark of bulrushes" in which the mother of Moses casts her baby adrift.

How did such correspondences pop up independently among cultures so separated in distance and in time as the people of ancient Mesopotamia and Palestine, and of nineteenth-century South America? We cannot argue that any constellations just "naturally" suggest "wickerwork" in their appearance in the sky -- and yet here we see these very same patterns appearing again and again around the globe, as if descended from some common ancient source.

The story of Sigu and the brown monkey, and the basket that is removed in order to initiate the flood, is an important clue in helping us to confirm the identity of the constellation Ophiucus as a correspondent to the arks and baskets in these various flood-narratives and baby-cast-adrift narratives.

We have already seen in our previous discussions of the deity Hanuman of ancient India that there is abundant evidence to argue that the constellation Hercules does in fact appear in some Star Myths in the form of a monkey -- no doubt due to the constellation's distinctively long rear-leg, which can be seen as a monkey-tail, as well as the constellation's very square-shaped head, which can be seen as sporting a ruff or a beard or thick sideburns, just like the heads of many monkeys and apes.

If the mischievous brown monkey from the myths of the Akawaio corresponds to the constellation Hercules in the sky, then the basket placed on top of the stump by the benevolent Sigu in order to prevent the inundation of the world undoubtedly corresponds to the triangular "lid" atop the rectangular "stump" of the constellation Ophiucus, towards which the constellation Hercules can be seen to be reaching his lower outstretched arm:

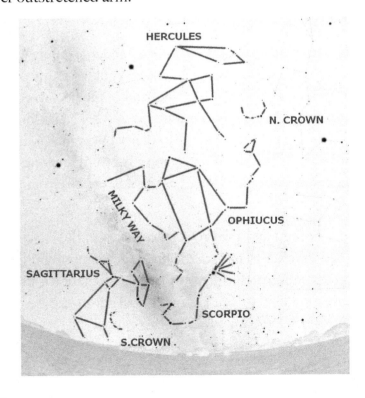

In the scene above, we see the monkey (played by the constellation Hercules) reaching down towards Ophiucus (the stump with the basket) and releasing the flood (likely the band of the Milky Way galaxy).

The Northern Crown may represent the basket after the monkey lifts it from the stump, and may also represent the basket with which the monkey was earlier assigned to carry water from the stream.

In fact, the flood myth appears to be one of the most oft-repeated myth patterns on earth, with flood legends appearing in the myths of cultures on every continent and inhabited island, and often with very similar patterns to those mentioned above. Many of these are discussed in *Hamlet's Mill*, in a chapter entitled "The Waters from the Deep" (the same chapter containing the account of Sigu and the troublesome monkey).

The flood myths almost always involve constellations adjacent to, or even partly "submerged" within, the shining band of the Milky Way galaxy, as are the constellations Ophiucus, Scorpio, Argo Navis, the Great Square of Pegasus, and Hercules.

I believe that the Mesopotamian versions of the flood-myth provide some of the best "windows" onto the possible meaning of these stories (and the related "baby-adrift" motif).

In the eleventh tablet containing the Gilgamesh cycle of myths, the semi-divine hero Gilgamesh has come to Uta-napishti the Distant in order to discover the secret of "the life eternal" (XI. 7). The account of the journey of Gilgamesh to Uta-napishti's abode actually takes place in Tablet X. Gilgamesh is distraught over the death of his companion Enkidu, and seeks out Uta-napishti in his grief, in hopes that the one who has gone through the flood can tell Gilgamesh the secret to overcoming the power of death.

When Gilgamesh arrives in the far land where Uta-napishti dwells, Uta-napishti asks Gilgamesh why his cheeks look so hollow and his face so sunken.

Gilgamesh replies by asking Uta-napishti how is it that he has stood with the gods in assembly -- because he perceives (to his surprise) that Uta-napishti is no different from Gilgamesh.

Then Uta-napishti says to him:

> "Let me disclose, O Gilgamesh, a matter most secret.
> To you I will tell a mystery of gods." XI. 9-10.[54]

Then Uta-napishti relates the entire account of the great deluge, concluding with the offering which Uta-napishti makes to the four winds at the conclusion of the flood. The text tells us that this sacrifice produces a sweet savor (a very strong parallel to the description in the Genesis account of the sacrifice of thanksgiving offered by Noah at the conclusion of the flood), after which the god Enlil comes to Uta-napishti and his wife, and declares that:

> "In the past Uta-napishti was a mortal man,
> but now he and his wife shall become like us gods!
> Uta-napishti shall dwell far away, where the rivers flow forth!" XI. 199 - 206.[55]

And here we find the key to the interpretation of this family of myth-patterns: Uta-napishti and his wife achieve "the life eternal," the "mystery of gods," by going *through the flood!*

And, as we have seen from the celestial correspondences outlined above, the flood through which Uta-napishti has passed is not a literal, historical, terrestrial flood -- it is an allegorical flood, based upon the system of celestial metaphor, and it is intended to teach us truths about an entirely different flood than a mythical flood which supposedly occurred many thousands of years ago.

I would argue that the story of the flood of Uta-napishti, and of Noah, and of Deucalion, and of Sigu, and of all the other flood myths around the globe, is intended to describe a flood that each and every one of us is going through right now, in this incarnate life.

It is the very same "flood" into which Odysseus is thrown, in the Odyssey of ancient Greece, when he is cast into the ocean by the wrath of Poseidon, and goes through a series of "spin cycles" which relate to the very same "ocean crossing" that Uta-napishti and his wife undertake in order to achieve the blessing pronounced by Enlil, that they shall "become like gods."

The message of Uta-napishti to Gilgamesh -- and to us -- is that this crossing of the flood (in which we are all presently engaged) has a very important purpose, and one which we should not be neglecting.

The purpose is *not* to gain immortality here in this physical body. We have already seen, in our examination of the "failed baptism" motif, that our plunge down in to this incarnate realm must inevitably end with the death of the body. The Gilgamesh cycle illustrates the same truth: the remainder of Tablet XI details Gilgamesh's unsuccessful attempts to gain a plant of immortality, as well as his unsuccessful attempts to go without sleep (which Shakespeare calls "death's counterfeit," in *Macbeth* II. iii, as well as in a similar allusion found in *Hamlet*).

Instead, I believe the message is that we should be working on transcending the "spin cycle" of this incarnate life, and on achieving the situation described in the metaphorical language of the myths as *finding the life eternal, standing with the gods in assembly, becoming as the gods*, and other ways of describing the integration with the divine realm.

I am convinced that the myths themselves were given to humanity for the purpose of assisting us in that very integration -- as were other ancient disciplines which have survived in some cultures which appear to have as their goal the same sort of elevation and transformation, including Yoga, Tai Chi, Daoist internal alchemy, qigong, and related forms of internal energy work, Tantric disciplines, shamanic practices, and the many types of meditation which are practiced around the world.

In all of the flood myths, and in the "baby-in-a-basket" myths, the motion of the story is "upwards along the Milky Way," representative of the raising of the divine nature, the kundalini force, the Christ within – the re-establishing of the Djed which was cast down.

In the Moses story in particular, as shown in *Star Myths of the World, Volume Three*, the character of Moses who begins as a babe who was set adrift at the "base" of the Milky Way column can be shown to take on the character of other constellations, moving upwards along the Milky Way itself.

These myths are very clearly dealing with questions about the purpose and meaning of this incarnate life, and with the inevitability of death -- an explicit theme in many of the flood myths, and in the quest of Gilgamesh to visit Uta-napishti after the death of Enkidu.

The message that they give us, in regard to these questions, is one of great hopefulness: it is by going through *this mortal life*, this "flood," that we can stand in the "assembly of the gods."

The Finding of Moses, Gustave Doré

Bringing a smile to the lips of the Goddess

In Genesis chapter 18, Abraham and his wife Sarah are visited by the LORD, and promised a child. The text of this famous passage tells us:

> 10 And he said, I will certainly return unto thee according to the time of life; and, lo, Sarah thy wife shall have a son. And Sarah heard *it* in the tent door, which *was* behind him.
> 11 Now Abraham and Sarah *were* old *and* well stricken in age; *and* it ceased to be with Sarah after the manner of women.
> 12 Therefore Sarah laughed within herself, saying, After I am old shall I have pleasure, my lord being old also?
> 13 And the LORD said unto Abraham, Wherefore did Sarah laugh, saying, Shall I of a surety bear a child, which am old?
> 14 Is any thing too hard for the LORD? At the time appointed I will return unto thee, according to the time of life, and Sarah shall have a son.
> 15 Then Sarah denied, saying, I laughed not; for she was afraid. And he said, Nay; but thou didst laugh.

This passage displays a myth-pattern we could call the "reluctant laugh" or the "reluctant smile," and is one we will see in other myths from other cultures.

In each case, the one who smiles is a woman, and in each case, there is a hint that the laugh has to do with something sexual in nature (in the Genesis passage cited above, we are specifically told that Sarah laughs within herself at the thought that she and her husband Abraham should "have pleasure" at their advanced age).

The pattern is found in the Norse myth of the beautiful Skade (or Skadi), goddess of skiers, and daughter of a powerful jotun named Tjasse (sometimes spelled Thiassi). Tjasse was killed by the Aesir in an incident involving the theft of the apples of youth, which the Aesir gods consumed in order to maintain their youth and power, and when Tjasse's daughter Skade learned of her father's death, she stormed into Asgard to avenge him. The account is related by Snorri Sturluson in the Skalskaparmal, part of the prose Edda:

> But Skadi, daughter of giant Thiassi, took helmet and mail-coat and all weapons of war and went to Asgard to avenge her father. But the Aesir offered her atonement and compensation, the first item of which was that she was to choose herself a husband out of the Aesir and choose by the feet and see nothing else of them. Then she saw one person's feet that were exceptionally beautiful and said:
>
> 'I choose that one; there can be little that is ugly about Baldr.'
>
> But it was Niord of Noatun.
>
> It was also in her terms of settlement that the Aesir were to do something that she thought they would not be able to, that was to make her laugh. Then Loki did as follows: he tied a cord round the beard of a certain nanny-goat and the other end round his testicles, and they drew each other back and forth and both squealed loudly. Then Loki let himself drop into Skadi's lap, and she laughed. Then the atonement with her on the part of the Aesir was complete.[56]

We find another account involving a "reluctant laugh" in the myths of ancient Greece, and once again the story involves a bereaved goddess and an obscene dance of some sort.

When Persephone was abducted by the god of the underworld, Persephone's mother Demeter searched the face of the earth for any clue regarding her daughter's disappearance. Just as Skade was certain that, after the killing of her father, it would be impossible to make her laugh, so also Demeter, after the abduction of her child, was in no mood for laughing -- but the myths relate that she was also induced to laugh.

The ancient author known as Pseudo-Apollodorus (because his work was originally thought to have represented the myth-collections of a writer by that name from the second century BC, until scholars pointed out that his writings reference sources from the first century AD) recounts that, when Hades carried off Persephone:

> Demeter went about seeking her all over the earth with torches by night and day, and learning from the people of Hermion that Pluto had carried her off, she was wroth with the gods and quitted heaven, and came in the likeness of a woman to Eleusis. And first she sat down on the rock which has been named Laughless after her, beside what is called the Well of the Fair Dances; thereupon she made her way to Celeus, who at that time reigned over the Eleusinians. Some women were in the house, and when they bade her sit down beside them, a certain old crone, Iambe, joked the goddess and made her smile. For that reason they say that the women break jests at the Thesmophoria.[57]

That these "jests" that make the goddess smile are obscene in nature, in fitting with the pattern found around the world, is evident from another ancient source, that of the famous ancient historian Diodorus Siculus (thought to have lived and written during the first century BC or BCE).

Diodorus tells us (speaking of the inhabitants of Sicily, where Demeter, according to ancient tradition, first shared the gift of wheat with humanity):

> In the case of Corê [that is, "the Maiden," an ancient way of referring to Persephone], for instance, they established

the celebration of her return at about the time when the fruit of the corn was found to come to maturity, and they celebrate this sacrifice and festive gathering with such strictness of observance and such zeal as we should reasonably expect those men to show who are returning thanks for having been selected before all mankind for the greatest possible gift; but in the case of Demeter they preferred that time for the sacrifice when the sowing of the corn is first begun, and for a period of ten days they hold a festive gathering which bears the name of this goddess and is most magnificent by reason of the brilliance of their preparation for it, while in the observance of it they imitate the ancient manner of life. And it is their custom during these days to indulge in coarse language as they associate one with another, the reason being that by such coarseness the goddess, grieved though she was at the Rape of Corê, burst into laughter.[58]

From the name of Iambe is thought to derive the name of the iambic verse, because (according to some sources) the jests and dances with which Iambe caused Demeter to smile were framed in that form of poetic verse.

It is significant that Iambe, who makes the goddess smile, is also associated with the figure known as Baubo, who is similarly described as an old woman who made Demeter smile. According to the Eleusinian tradition, Baubo made Demeter smile not only through lascivious talk and dancing, but also by exposing her genitalia. Clearly, a consistent pattern in these myths of the "reluctant smile" is emerging which involves sexuality.

It is also notable that Pseudo-Apollodorus describes Iambe as performing a "failed baptism" very reminiscent of the activities of Isis at Byblos. Immediately after explaining that the jests of Iambe caused Demeter to laugh, the ancient author tells us:

> But Metanira, wife of Celeus [in whose kingdom Demeter was searching for news about Persephone, and in whose palace Iambe made Demeter laugh] had a child and Demeter received it to nurse, and wishing to make it immortal she set the babe of nights on fire and stripped off its mortal flesh. But as Demophon -- for that was the child's name -- grew marvelously by day, Praxithea watched, and discovering him buried in the fire she cried out; wherefore the babe was consumed by the fire and the goddess revealed herself.[59]

A similar account involving the goddess Demeter herself is found in the second "Homeric Hymn" of ancient Greece. Truly, the evidence that myths from around the world somehow share a common ancient source can be accurately said to be overwhelming and conclusive!

Still further examples of the pattern of the "obscene dance associated with laughter" can be found in Star Myths from different cultures. One of the most famous episodes in the Kojiki of ancient Japan involves the retreat of the goddess Amaterasu, the "Heavenly-Shining-Goddess," into a rock cave due to the outrageous antics of her brother Susanowo, whose name means "Brave Swift Impetuous Male." The celestial foundations of this important episode are examined in *Star Myths of the World, Volume One*.[60]

After Susanowo drops a "backward flayed" heavenly piebald horse through a hole in the top of the sacred weaving-hall where Amaterasu is supervising the weaving of the heavenly garments by her attendants, we are told that the goddess is so upset that she retreats into her Heavenly Rock-Dwelling and closes the door behind her, plunging the Plain of High Heaven and all the Central Land of Reed-Plains into deep darkness.

The Kojiki informs us that, seeing the world covered in unending night, the eight hundred thousand deities assembled in the bed of the Tranquil River of Heaven to try to find a way to coax Amaterasu back out of the cave. We are told that their plan includes:

> assembling the long-singing birds of eternal night and making them sing, taking the hard rocks of Heaven from the river-bed of the Tranquil River of Heaven, and taking the iron from the Heavenly Metal-Mountains, calling in the smith Ama-tsu-ma-ra, charging her Augustness I-shi-ko-ri-do-me to make a mirror, and charging His Augustness jewel-Ancestor to make an augustly complete [string] of curved jewels eight feet [long], -- of five hundred jewels, -- and summoning His Augustness Heavenly-Beckoning-Ancestor-Lord and His Augustness Great-Jewel, and causing them to pull out with a complete pulling the shoulder [-blade] of a true stag from the Heavenly Mount Kagu, and take cherrybark from the Heavenly Mount Kagu, and perform divination, and pulling up by its roots a true *cleyera japonica* with five hundred [branches] from the Heavenly Mount Kagu, and taking and putting upon its upper branches the augustly complete [string] of curved jewels eight feet [long], -- of five hundred jewels, -- and taking and tying to the middle branches the mirror eight feet [long], and taking and hanging upon its lower branches the white pacificatory offerings and the blue pacificatory offerings, His Augustness Grand-jewel taking these divers things and holding them together with the grand august Offerings, and His Augustness Heavenly-Beckoning-Ancestor-Lord prayerfully reciting grand liturgies, and the Heavenly Hand-Strength-Male-Deity standing hidden beside the door, and Her Augustness Heavenly-Alarming Female hanging [round her] the heavenly clubmoss of the Heavenly Mount Kagu as a sash, and making the heavenly spindle-tree her head-dress, and binding the leaves of the bamboo-grass of the Heavenly Mount Kagu in a posy for her hands, and laying a soundingboard before the door of the Heavenly Rock-Dwelling, and stamping till she made it resound and doing as if possessed by a Deity, and pulling out the nipples of her breasts, pushing down her skirt-string all the way to her private parts. Then the Plain of High Heaven shook, and the eight hundred myriad Deities laughed together. Hereupon the Heaven-Shining-Great-August-Deity [Amaterasu] was amazed, and, slightly opening the door of the Heavenly Rock-Dwelling, spoke thus from the inside: "Methought that owing to my retirement the Plain of Heaven would be dark, and likewise the

> Central Land of Reed-Plains would be all dark: how then is it that the Heavenly-Alarming-Female [Ame-no-Uzume-no-mikoto] makes merry, and that likewise the eight hundred myriad Deities all laugh?"[61]

Once again, we have a female deity who is in no mood for laughter, and we have an obscene dance which arouses her interest -- although in this myth from ancient Japan, it is the eight hundred thousand deities and spirit beings who shake the heavens with their laughter at the antics of the alarming female.

All of the various components of these myths, comprising a pattern which can be seen to share certain common elements, even though the details of each manifestation are very different, can be shown to be based upon specific constellations.

In all of these myths in which a goddess or other important female personage must be made to smile, and in which the smile is ultimately elicited via sexual talk, sexual thoughts, or lewd dancing, or the exposure of genitalia, I believe it is abundantly evident that we are dealing with the constellation Virgo the Virgin in the night sky -- a constellation which indeed can be seen to wear a smile, if viewed on a clear night from a location far away from light pollution, as illustrated on the following page.

Examine the close-up chart showing the zodiac constellation Virgo on the opposite page, in order to verify that the outline of her form does indeed contain stars which can be seen to form a smile. Next, examine the star chart below to see Virgo in context with other nearby constellations that feature in these myths:

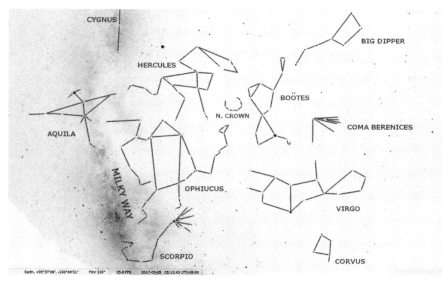

Cygnus (not fully visible, at top) and Aquila represent the "long-singing birds of eternal night" in the Kojiki account. The "Tranquil River of Heaven" is the Milky Way band.

The constellation Ophiucus, which in other Star Myths can be shown to play the role of a door or a gate, probably represents the "tent door" within which Sarah hears the promise of a child in the Genesis account, when she laughs, as well as the door to the "Heavenly Rock-Dwelling" of Amaterasu.

The Northern Crown plays the role of the curved necklace of jewels, eight feet long (note there are eight stars connected in the illustration).

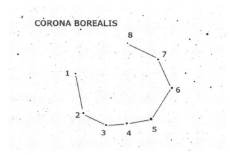

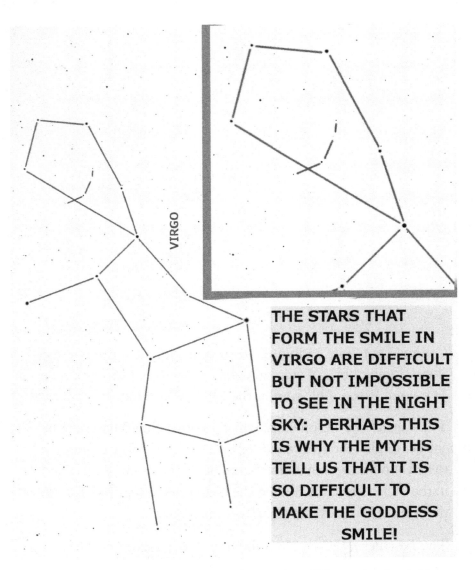

THE STARS THAT FORM THE SMILE IN VIRGO ARE DIFFICULT BUT NOT IMPOSSIBLE TO SEE IN THE NIGHT SKY: PERHAPS THIS IS WHY THE MYTHS TELL US THAT IT IS SO DIFFICULT TO MAKE THE GODDESS SMILE!

In the above image, showing the smile of Virgo, it should be noted that while the outline of the constellation has been drawn in accordance with that envisioned by H. A. Rey in his book *The Stars: A New Way to See Them*, Rey himself never (to my knowledge) mentioned the smile or drew it in to his outlines of the constellation Virgo.

To return to the surrounding constellations which play a role in some versions of the story, the very faint but mythologically important constellation Coma Berenices (or "Berenice's Hair") can be seen directly above the "outstretched arm" of the constellation Virgo. This constellation features in numerous myths involving Virgo-figures -- sometimes in the role of hair (often hair which has been cut off, such as in the story of Samson in the Bible, or the story of the goddess Sif in the Norse myths), but also in the role of a "posy" of flowers, a "whisk" or fan of feathers or other material, and -- in the Kojiki account of Amaterasu cited above, as the "posy of bamboo grass" which Heavenly-Alarming-Female (Ame-no-Uzume) carries in her hand during her suggestive dancing.

Based upon this identification, as well as on the description of her actions in the dance itself (such as pushing her skirt-string down below her private parts), we can surmise that Heavenly-Alarming-Female is probably *also* played by the constellation Virgo in the Kojiki account (Amaterasu herself is not described as laughing in this particular version of the story -- instead, it is the "eight hundred myriad" assembled Kami who laugh, and who are probably represented by the countless stars of the Milky Way band, along with the other numberless stars of the heavens, as Heavenly-Alarming does her dance).

In other versions of the myth, such as the story of Skade, it is a male character who performs the off-color dance -- this time, the figure of Loki the trickster god, who ties himself to a goat and cavorts around squealing until he lands in the lap of the goddess and elicits her smile.

The character of Loki in this instance is almost certainly represented by the celestial figure of Boötes the Herdsman, who is positioned above Virgo and does indeed look to be on the verge of falling into her lap.

Boötes is an important constellation, and characters associated with this celestial figure play a role in many Star Myths from around the globe. The constellation contains the large reddish star Arcturus, which can be located quite easily by following the sweep of the handle of the Big Dipper, which guides the viewer along an arcing line from the tip of the handle to the star Arcturus – hence the saying "follow the arc to Arcturus."

The idea that Loki ties a certain sensitive part of his anatomy to a goat is probably explained by the fact that the Big Dipper plays a team of goats in Norse myth, and Loki is associated with the the figure of Boötes, who can be shown to play the role of a charioteer in numerous myths, including as Lord Krishna in the Mahabharata and the Bhagavad Gita of ancient India (discussed in *Star Myths of the World, Volume One*), and in the ancient Greek myths found in the Iliad (discussed in *Star Myths of the World, Volume Two*).

In these myths, the figure of Boötes plays the role of charioteer, and the stars of the Big Dipper represent the team of horses pulling the chariot. The reins between the charioteer and his team are represented by the stars of handle of the Dipper, and we can envision a connecting line between this handle and the "pipe" of Boötes, which could represent the extended hand of the charioteer.

In the Norse myths, the chariot of Thor is drawn by billy-goats -- and the figure of Thor himself can be identified with the powerful form of the constellation Hercules, which often plays the most powerful figure in any pantheon (and the figure with the most powerful weapon). These goats, of course, are represented by the Big Dipper -- and, in the legend in which Thor dresses up as the love-goddess Freya (or Freyja), Loki is the charioteer who drives the cart bearing Thor in disguise, thus indicating that Loki was associated with Boötes, and that the Dipper was associated with a team of goats.

Thus, although it may not seem obvious without much knowledge of the ancient myth-system, we can see that Loki's act of tying himself to a goat in order to make Skade laugh is a mythical scene which is almost certainly connected to Boötes and the Big Dipper.

In fact, the large and bulbous head of Boötes, with the extended "pipe" shaped line of stars, appears to have had some sexual connotations in the ancient worldwide myth-system. Figures associated with Boötes in myth are sometimes depicted as carrying a large gourd, and to have humorous sexual connotations.

For example, the image on the following page shows a male character named Daikoku, who holds over his head a large gourd which clearly has some resemblance to the male sexual organs. Just below him in the image dances his companion Otafuku, a female figure from Japanese mythology who is devoted to the very goddess Ame-no-Uzume (Heavenly-Alarming-Female) whose suggestive dancing coaxed Amaterasu from the cave. Otafuku, like Uzume herself, is a fertility figure who radiates open and frank sexuality.

Note that in the image, Otafuku is dancing with one hand on her hip and with the other extended and holding a long wand. Her posture is clearly illustrative of the constellation Virgo -- right down to the distance between her two feet and the angle of her body.

The extended arm corresponds to the extended arm of the constellation itself, which we have already noted to be one of the most distinctive features of the constellation Virgo, marked by the star Vindemiatrix.

These connections between the constellation Boötes and sexuality cannot be said to be obvious, and yet it is clear that the connection appears in myths from around the world, separated by vast distances.

But what could this myth-pattern of the "reluctant laugh" possibly be trying to convey to our understanding? The pattern itself is obviously very prevalent in world myth, and yet its meaning is not at all easy to decipher.

We can perhaps best unravel their esoteric meaning by examining the rituals surrounding the Eleusinian Mysteries of the ancient world, which took place every year in ancient Attica for many centuries, from at least the sixth century BC and probably even earlier, until they were finally shut down by decree of the emperor Theodosius in AD 392, during the rise of literalist Christianity.

Although the secrets of the Eleusinian Mysteries were well guarded, some of the rituals leading up to the climactic culmination of the mysteria are known from ancient sources, including Aristophanes, Herodotus, Strabo and Plutarch. Among these rituals was the procession of the initiates from Athens to Eleusis, during which the initiates were harassed and insulted by onlookers called *gephrysmoi*, and instructed to shout obscenities while crossing a certain stream (the Kephissos).[62] These ritual insults were associated with the role of Iambe in causing Demeter to laugh.

Other aspects of the preparatory rituals leading up to the climax of the Eleusinian Mysteries included the procession of the initiates to the waters of the sea, carrying a piglet. I would argue that this indicates that the rites of Eleusis involved the dramatization of the plunge into incarnate life which we have previously seen to be esoterically described as an immersion in the watery element representative of the lower realms, as well as the imprisonment of the divine soul within the animal form of the physical body.

This interpretation of the Eleusinian Mysteries is supported by the timing of the mysteria during the calendar of the year, which took place at a certain prescribed period of time following the autumnal equinox -- the point on the annual cycle which we have already seen to be representative of the plunge down from the realm of spirit into the realm of matter.

It may be that the myth-pattern involving laughter in the face of tragic circumstances (such as Demeter's loss of her daughter Persephone, or Skade's loss of her father Tjasse, or even Sarah's inability to have children and her realization that she is now well past the age of child-bearing) is somehow related to the realization that our incarnate condition is itself laughable.

The message to us may in some way relate to the fact that our possession of a body is an unusual -- and temporary -- circumstance. The fact that all of the smiles in these myths are elicited by an overt reference to the sexual aspect of our incarnate condition may in some way be an invitation to us to keep our temporary incarnation in an "animal" body in its proper perspective.

It may be that the *immersion in the sea*, while *carrying a piglet*, or the ritual mocking while *crossing a bridge* of the initiates into the Eleusinian mysteria were intended to provide the same sort of perspective. We are, in a sense, "crossing a bridge" in this incarnate life, or undergoing a difficult immersion in the "sea" of the lower elements, in an "animal" body.

The determination of Demeter or Skade to refuse to smile is of course understandable. Anger and grief at death or loss are not misplaced -- and the myths do not imply in any way that these responses and emotions are unwarranted.

But they do seem to be telling us that even in our proper anger and grief at the pain and loss and suffering that we experience in this lower realm, we can also -- as incongruous as it may seem -- laugh. Because this incarnate life is not meant to be a permanent condition (the presence of "failed baptism" stories in the Demeter myth testify to that: immortality in this incarnate life is never held up in myth as a goal).

There undoubtedly must be many other layers of meaning for us in the myths of the goddess or important female figure who must be made to smile.

But at the heart of the message seems to be the important perspective that the physical realm is not the most important -- even though (like Sarah in the Genesis story) we often get caught up in the exigencies of the ordinary realm, and forget where we came from, and overlook the *miraculous* origin of *any* human birth.

And not only that, but the rebirth of the grain and all the other plants that sustain the animal world, and hence all human life, must also be seen to be a miracle, and a gift from the Invisible Realm -- from the realm of the gods, and (in the terminology of the ancient Greeks) from Demeter and Persephone in particular.

The ancient myths seem to be showing us that we ourselves are part of the same type of cycle, descending from the realm of spirit into what *seems to us* to be the realm of matter, but only for a temporary "crossing of the bridge," and (the myths also seem to be telling us) in a way that is not really reflective of our true divine nature (in fact, our incarnate condition is somewhat laughable).

In fact, if we think about laughter in gerneral, laughter itself often involves a sudden, abrupt *shift* in perspective -- a sudden realization (such as in the punchline of a joke) that the situation is totally different from what was previously assumed or expected, or a sudden indecorous moment which causes everyone present to realize that things aren't as serious as everyone was previously imagining them to be.

These ancient myths involving the smile of the serious goddess seem to be intended to elicit the same sort of *sudden shift* in perspective in us, on a scale involving our human condition in this seemingly physical universe, and to get us to see that things aren't necessarily as grim as we might imagine them to be.

The Retrieval from the Dead

The story of the failed retrieval from the underworld finds so many parallels in the world's ancient myths and sacred stories that the connections cannot be ignored -- providing very strong evidence of a common origin. As we will see, the evidence for the celestial foundation of these similar myths is equally compelling.

Perhaps the most familiar of the "failed retrieval" stories to western audiences is the story of Orpheus and Eurydice from ancient Greece. Although the story appears in different versions (not uncommon in Greek myth, although this did not appear to bother anyone), in all of them the basic outline remains the same.

Orpheus of Thrace was the son of one of the nine Muses, Calliope, and divinely gifted with musical genius beyond any other mortal. His skill with the lyre and the beauty of his voice could charm even inanimate objects, including rocks and trees, to say nothing of animals and mortal men and women.

Orpheus loved the beautiful maiden Eurydice, but on the day of their marriage, she stepped on a deadly adder while dancing and died. In some variants of the myth, Eurydice was pursued by a lustful god and stepped on the serpent while fleeing. In either case, she perished from the effect of the snake's deadly venom, leaving Orpheus heartbroken.

Unwilling to accept her death, Orpheus determined to find his way to the underworld and beseech the ruler of the underworld -- and his queen, Persephone -- to relinquish Eurydice and allow her to return to the land of the living.

The plea of the matchless singer, full of pathos regarding his unsuccessful attempts to overcome his grief, as well as his respectful deference to the god and goddess who rule over the realm of the dead, and the power of his tuneful harp, moved the rulers of that dread kingdom to release Eurydice and allow her to follow Orpheus back above – but only on the condition that he *must not look back* at all until he had completely exited the underworld.

The ancient poets describe how the hopeful duo climbed the dark, cold, fog-shrouded path, leading steeply upwards through the far reaches of the land of the dead, towards the very boundaries of the upper world, the realm of the living – but that here, Orpheus was overcome by the need to make sure she was going to make it out, and he turned his gaze backwards for the first time, to look at his beloved.

All he saw was a glimpse, as she slipped back into the land of shades.

Orpheus reached out to try catch her, but his arms embraced nothing but empty air.

And this is the unfailing pattern of the myth of the *retrieval from the underworld*. In nearly every case, the condition is the same: no looking back. And, in nearly every case, the outcome is also the same: at the last possible moment, when success seems certain, the temptation to look overwhelms the rescuer – and the beloved slips back into the gloom, never to return.

We see a very similar story involving the unsuccessful retrieval of the beloved from the underworld in many other myths around the world – including in cultures which would seem to have had little opportunity for contact with the source of the ancient Greek version, such as the myths of the Maya of Central America, or in the Kojiki of ancient Japan. In every case, the outline of the myth is nearly identical.

In the version found in the Kojiki, for instance, it is the progenitor god and goddess, Izanagi and Izanami (whom we first met in a previous chapter), who act out the pattern. The Kojiki informs us that Izanami is so badly burned while giving birth to a fiery kami that she succumbs. Just as in the myth of Orpheus, Izanagi determines to take the grim journey to the underworld, in order to try to obtain the release of his beloved wife.

Beginning in section seven of the Kojiki, we learn that in giving birth to the deity Fire-Shining Prince, Izanami is severely burned, and that she lies down afterwards in great pain and eventually perishes, while Izanagi laments and weeps. Then, he buries her "on Mount Hiba at the boundary of the Land of Idzumo and the Land of Hahaki."[63]

In the translation of Basil Hall Chamberlain (1850 - 1935), the story of Izanagi's attempt to bring Izanami back from the underworld is told in the Kojiki thus:

> Thereupon Izanagi, wishing to see his younger sister Her Augustness the Female-Who-Invites, followed after her to the Land of Hades [literally *Yomo* or *Yomi*, the "Yellow Stream"]. So when from the palace she raised the door and came out to meet him, His Augustness the Male-Who-Invites spoke, saying: "Thine Augustness my lovely younger sister! the lands that I and thou made are not yet finished making; so come back!" Then Her Augustness the Female-Who-Invites answered saying:
>
> "Lamentable indeed that thou camest not sooner! I have eaten of the furnace [or cooking fire] of Hades. Nevertheless, as I reverence the entry here of Thine Augustness my lovely elder brother, I wish to return. Moreover, I will discuss it particularly with the Deities of Hades [*Yomo-tsu-kami*]. Look not at me!" Having thus spoken, she

161

went back inside the palace; and as she tarried there very long, he could not wait. So having taken and broken off one of the end-teeth of the multitudinous and close-toothed comb stuck in the august left bunch [of his hair], he lit one end and looked. Maggots were swarming, and [she was] rotting, and in her head dwelt the Great-Thunder, in her breast dwelt the Fire-Thunder, in her right hand dwelt the Earth-Thunder, in her left foot dwelt the Rumbling-Thunder, in her right foot dwelt the Couchant-Thunder: – altogether eight Thunder-Deities had been born and dwelt there. Hereupon His Augustness the Male-Who-Invites, overawed at the sight, fled back, whereupon his younger sister Her Augustness the Female-Who-Invites said: "Thou hast put me to shame," and at once sent the Ugly-Female-of-Hades [*Yomo-tsu-shiko-me*] to pursue him. So His Augustness the Male-Who-Invites took his black august head-dress and cast it down, and it instantly turned into grapes. While she picked them up and ate them, he fled on; but as she still pursued him, he took and broke the multitudinous and close-toothed comb in the right bunch [of his hair] and cast it down, and it instantly turned into bamboo-sprouts. While she pulled them up and ate them, he fled on. Again later [his Younger sister] sent the eight Thunder-Deities with a thousand and five hundred warriors of Hades to pursue him. So he, drawing the ten-grasp sabre that was augustly girded on him, fled forward brandishing it in his back hand; and as they still pursued, he took, on reaching the base of the Even Pass of Hades [Or Flat Hill of Hades, *Yomo-tsu-hira-saka*], three peaches that were growing at its base, and waited and smote [his pursuers therewith] so that they all fled back. Then His Augustness the Male-Who-Invites announced to the peaches: "Like as ye have helped me, so must ye help all living people in the Central Land of Reed-Plains when they shall fall into troublous circumstances and be harassed!" -- and he gave [to the peaches] the designation of Their Augustness Great-Divine-Fruit. Last of all his younger sister Her Augustness the Princess-Who-Invites came out herself in pursuit. So he drew a thousand draught rock, and [with it] blocked up the Even Pass of Hades, and placed the rock in the middle; and as they stood opposite to one another and exchanged leave takings [. . .].[64]

This pattern can be seen to have echoes which reverberate through many other similar myths and folktales from around the world. The pattern of the one being pursued throwing down items which the terrifying pursuer stops to collect or consume is especially prevalent.

The detail of the one who has gone down to the underworld being unable to leave, even though she wants to, because she has consumed the food of the land of the dead is also very well-known in the different versions of this myth-pattern: for instance, in the Persephone myth, Persephone is released to return to the land of the living, but it is revealed that she has tasted the food of the underworld (in most versions, a certain number of pomegranate seeds) and thus must remain in Hades for a certain number of months each year.

Also frequently seen in versions of this story is the detail of the one who has eaten in the underworld, and who thus must remain in that realm, being seen to be corrupted, putrefying, and eaten by worms or maggots. We see this same detail in some of the similar myths from the indigenous cultures of the Americas, whose sacred traditions often include a version of the "failed retrieval from the dead."

Indeed, in the sacred stories of the indigenous cultures of the Americas we encounter a veritable treasure-trove of myths involving the retrieval from the land of the dead, including a great number of variations upon the failed-retrieval pattern seen in the Orpheus and Eurydice myth of ancient Greece and the Izanagi and Izanami myth of ancient Japan.

Writing in the first half of the twentieth century, and citing first-hand accounts of narrators from various nations of the Native American people, Dr. Anna H. Gayton published an extensive study entitled "The Orpheus Myth in North America" in the *Journal of American Folk-Lore* in the summer of 1935.

There, she documents evidence of the pattern stretching across the continent, from the Chinook, Nutka, Tlingit and Kwakiutl in the northwest, to the Modoc, Shasta, Miwok, Mono, and Yokuts in the west, to the Navaho and Zuñi in the southwest, to the Blackfoot, Pawnee, Wichita and Tawakoni in the central plains regions, to the Koasati, Alibamu, Cherokee and Yuchi in the southeast, and the Huron, Seneca, Winnebago, Fox, Menomini, Mantagnais, Malecite, and Micmac further north along the eastern reaches of North America (among others; all spelling as published in the map accompanying the original article).[65]

The unifying pattern usually involves several elements, which vary from one manifestation to another, including: the following of the departed beloved to the other world, involving travel usually towards the west, receiving aid from a supernatural helper, surmounting terrifying obstacles (often involving crossing a rushing stream by way of an unstable bridge, as well as other physical obstacles), an encounter with the shades of the departed in the land of the dead whose primary activity appears to be dancing in a ritual manner, discouragement from the deceased who does not want to leave the Other Realm, assent and cooperation from an authority figure in the realm of the dead, the consumption of food in the underworld or the temptation to consume food there, the recovery of the deceased soul (sometimes in a container of some sort), and the contingent conditions which must be observed in order to enable the successful return of the departed, which are usually violated (resulting in the final loss of the departed beloved) but successfully observed in some less common examples.[66]

A thorough examination of each manifestation – and the strong connections to specific constellations which I believe these stories exhibit – would fill an entire volume, and one that would be very beneficial to undertake in the future. However, such a detailed study is beyond the scope of this particular chapter, and so only a few variants will be mentioned in order to establish beyond a doubt that in this pattern of the "failed retrieval" we have one of the strongest pieces of evidence for the operation of a worldwide system of celestial metaphor, and one that can hardly be denied so well documented and so abundant are the examples.

The first example Dr. Gayton cites was related to her during the year 1929, and comes from the Telumni Yokuts people of western North America. As in most examples of this myth-pattern, including that of Orpheus and Eurydice, a beloved wife has died and her husband decides to follow her to the underworld. In many of the Native American examples, the deceased is a beloved sister, which also resonates with the myth as found in the Kojiki of Japan.

A very interesting aspect of the Yokut version involves the crossing of a hazardous bridge over a river, which is a feature in many of the other versions collected by Dr. Gayton (some Native American versions also involve a ferocious dog as well as a river: and note that both of these elements are well-known features of the myths of ancient Greece, in which the land of the underworld is guarded by Cerberus and is reached by crossing the River Styx).

The shade of the wife tries to tell her husband, who is following her towards the west, that he cannot cross this bridge, because he is alive -- he will fall into the river and be turned into a fish. The narrator also tells us that:

> In the middle of the river is a bird, kildeer, who tries to scare those who are crossing by suddenly saying, "Kat, kat,

kat!" If a person loses his balance and falls in the water he becomes a fish forever.[67]

The husband, however, is able to skim safely across the bridge by virtue of a magic "eagle down rope." However, when he reaches the land of the dead, he fails in the tests that are given to him by the chief of that place – primarily involving the requirement to remain awake, which he consistently fails to do. In that version of the story, he returns without his wife, and is warned that he must not relate what happened until after six days had passed, or he himself will die -- a test he also fails.

The conditions that must not be violated vary from one culture to another. In some versions, the wife comes back from the land of the dead, but the husband may not touch her or embrace her for a certain period of time -- and when he fails to restrain himself, she vanishes (or is replaced by a rotten log).

In a Navaho (or Navajo) version cited in the essay, the husband is told not to build a fire while he is in the underworld: he violates these instructions and sees that the dead he had seen dancing the night before were now a field of skeletons strewn over the landscape in tattered blankets -- and he flees, pursued by his dead wife. When they reach the land of the living once more, some animals offer to help restore her to life, by performing a ceremony during which the man must not look at his wife. Predictably, he violates these instructions, and finds that she is now a skeleton.[68]

A few other noteworthy elements found in the first-hand accounts recorded in Gayton's article are a tempting strawberry, which the husband must resist the urge to devour, the dancing of the dead in a circle each night (which in some cases involves the healing of the earth), a hut or lodge in which the husband stays while he is in the realm of the dead, assistance from an old woman who is called "Wind" in some versions of the story, and in many cases a gourd or box in which the soul of the departed can be brought back to the land of the living, as long as it is not opened (a curious bystander usually peeks in to the gourd or the box once the husband has successfully returned, foiling the return of the departed beloved -- recalling the story of Sigu and the curious brown monkey).

One of the earliest recorded examples of this "failed retrieval" pattern among the indigenous nations of the Americas was recorded among the Huron nation by a Jesuit missionary, Jean de Brébeuf (1593 - 1649). He cites it as one of two examples of what he (or his translator) condescendingly calls "two of the most stupid" of the "stories which the fathers tell their children."[69]

In the accounts he records, he relates that the Village of the Dead is reached by way of a road crossed by a rushing river with a narrow bridge, guarded by a dog which jumps at many souls trying to cross, causing them to fall into the torrent. Before arriving at the Village of the Dead, one first encounters the cabin of a being named Oscotarach, or "Pierce-Head," who draws the brains out of the heads of the deceased, and keeps them.

In the specific Huron myth of the man who follows his departed sister to the underworld, in order to bring her back, the adventurer encounters the one who keeps the brains of the dead, in his cabin, and receives from him a pumpkin into which he can put her soul. He is able to tackle her as she dances among the other spirits, and eventually forces her into the pumpkin. On his way back, the one in the cabin gives him another pumpkin, into which he has put her brain, so that the brother can perform a ceremony to restore his sister, once he returns to their home. The ceremony is to be performed at a feast, during which all the guests must keep their eyes averted, as the brother carries both pumpkins through his cabin, until he gets to his own place at the table and sits down, after which the sister's body will be restored to life.

All goes well during the ceremony, until the brother is just a few steps from completion – when one curious guest raises his eyes to see what the ceremony involves. At that moment, the soul escapes and returns to the Village of the Dead.[70]

Anna H. Gayton in her article notes this account from Brébeuf as evidence that this myth, so similar in many ways to the Orpheus myth, was present in the Americas "at a time prior to any but the slightest European intrusion."[71] She also points out that these Native American versions could not have been seeded by the European intruders themselves, because many aspects of the story of Orpheus and Eurydice are *not* present, while many other different and unique details *are* present.[72] In other words, the myths differ greatly in their surface details, and yet it is very evident that the underlying pattern is strikingly familiar from one culture to the next, whether we examine the variants in the Americas, or in Japan, or in ancient Greece and Rome.

Later, building on the observations of Dr. Gayton, another scholar of comparative religion with a focus on Native American sacred tradition, Professor Åke Hultkrantz of Sweden, collected additional examples of this same pattern, publishing a book entitled *The North American Indian Orpheus Tradition* (1956).

In one notable example recorded by Hultkrantz, from a first-hand Comanche account recorded in 1933, the beloved wife dies soon after the marriage (very similar to the Orpheus story), the husband goes to the underworld and is told by the girl's father that she can return with him, and will be restored to life if upon arriving in the land of the living, the husband gives her a buffalo kidney to eat. The condition established for the successful retrieval from the underworld is that the husband may not strike his wife, and if he ever does so she will return to the realm of the dead.[73]

After they return to the land of the living, the kidney is consumed and the couple are happily reunited – but one night, he wishes to pull their buffalo blanket over the two of them in order to embrace her, and as he pulls on the blanket, his hand slips and strikes his wife. He is horrified, as with a fading cry she vanishes again into the Other World.

All of these stories, whether from ancient Greece or ancient Japan or from one of the many indigenous nations of North America, can be shown to have very specific connections to constellations in the night sky. Importantly, some of the details related above from the many Native American versions relate to different constellations than those forming the basis for the Orpheus story, for example (but all appear to be based upon the same general region of the sky).

This suggests that the pattern was not "borrowed" from one culture or another, but rather that we may be dealing with a very ancient myth-pattern that was altered down through the ages by people who still understood the connection to the constellations. Let us briefly examine the celestial players in this widespread myth, and see how their distinctive constellational characteristics give rise to the many different elements we've noted in the versions cited above.

The region of the sky involved in all these "retrieval from the underworld" myths surrounds the bright portion of the Milky Way band between Scorpio and Sagittarius -- the brightest and widest portion of the Milky Way. Just above Scorpio we see Ophiucus, a figure which we have seen to play the role of the "gates of the underworld" in the myths of ancient Greece because of its distinctive "doorway" shape (see discussion in *Star Myths of the World, Volume Two*).

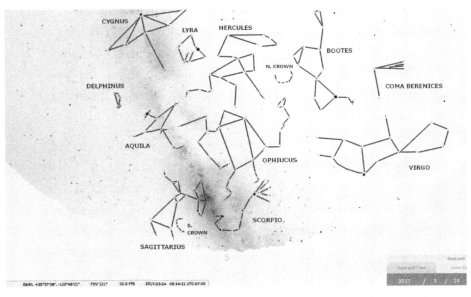

Many elements of the retrieval myth line up quite clearly with a very specific constellation, although some are ambiguous enough to allow a few possible interpretations.

In the first example presented, that of Orpheus and Eurydice, the beautiful maiden is slain when she treads on a serpent -- often a mythical embodiment of the position of Ophiucus directly above Scorpio. It is also possible that Virgo, positioned directly above a different celestial serpent (Hydra, not visible in the chart above because most of it has disappeared beneath the horizon on the west or right-side of the chart, just below Virgo), represents the beloved wife after she has been bitten, although the specificity of the story in which Eurydice is always bitten in the area of the foot argues persuasively for identification with Ophiucus above Scorpio.

Orpheus himself is most likely identified with the constellation Hercules, which is located immediately adjacent to the constellation Lyra, the Lyre, in the night sky. We saw in *Star Myths of the Bible* that the figure of David in the Hebrew Scriptures is most-frequently associated with the figure of Hercules as well -- and David was also described as skilled at the lyre or harp, just as Orpheus is described.

However, when Eurydice dies, Orpheus descends to the underworld in order to try to persuade the god and goddess of death to allow her to return. Look again at the outline of the constellation Sagittarius: can you see what aspects of the constellation give rise to the famous aspect of the story, in which *looking back* causes Orpheus to lose his beloved?

I am convinced that when he is allowed lead Eurydice forth from the underworld, but is strictly enjoined that he must not look back, Orpheus has descended to the location of Sagittarius. The outline of Sagittarius clearly appears to be *walking in one direction* (towards the east, or left side of the star chart above) while *looking back over its shoulder* towards the Milky Way (towards the right of the star chart, or towards the west).

In the version of the unsuccessful retrieval found in the Kojiki of Japan, Izanami herself admonishes Izanagi not to look at her. When he becomes impatient and does look, she is revealed to be decomposing and swarming with maggots or worms: this description is very typical of Scorpio figures in many Star Myths of the world. Note that if Sagittarius is envisioned as "looking back" as it proceeds to the east, the constellation that it would be imagined to be looking upon would indeed be Scorpio.

Intriguingly, it seems that many more celestial details in this "underworld retrieval" oicotype have been preserved among the various nations and peoples of the Americas than in either the Orpheus tradition of Greece or the Kojiki tradition of Japan. This richness of preserved celestial detail is evident from the few Native American versions of the pattern that have been cited above – and if we were to delve into all of the examples cited in the work of Gayton and of Hultkrantz, we would see still more celestial imagery which has been preserved in the different variations found among the sacred traditions of the Native American nations.

In the interest of space and brevity, the specific mythical details found in the versions cited above will be addressed below in a "list format," along with my interpretation of the constellational or celestial details which form the basis for those aspects of the stories:

The torrential river with the narrow, unsteady, dangerous bridge: The Milky Way band, and specifically the widest, brightest part of the Milky Way which runs between Scorpio and Sagittarius and also alongside the form of Ophiucus. Note that in this wide and bright portion of the galactic band (which actually corresponds to the Galactic Core), there is a dark pathway, known as the Great Rift or the Galactic Rift. This dark, narrow, winding channel across the brightest part of the Milky Way itself almost certainly corresponds to the "unsteady bridge" which is present in so many of the Native American versions of the underworld retrieval myth. This path or "bridge" runs between the arrows shown in the image below, and I am quite convinced it is the same path by which Moses leads the people across the Red Sea in the Hebrew Scriptures:

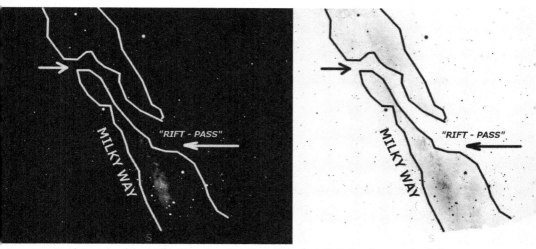

Both images of the Milky Way show location of the dark "Rift-Pass" across the Galaxy. Note that in the color-inverted version, on the right, everything that appears light is actually dark in the night sky, while area which appears darker represents the shining band of the Milky Way.

The admonition that those who fall into the river will turn into fish: The constellation Delphinus, the Dolphin, which "swims" or leaps alongside the torrential river of the Milky Way, and which very much resembles a leaping dolphin (or salmon) when seen in the night sky. The star chart from three pages previous is reproduced below for ease of reference; note Delphinus, swimming alongside the Milky Way, between the figures of Cygnus and Aquila.

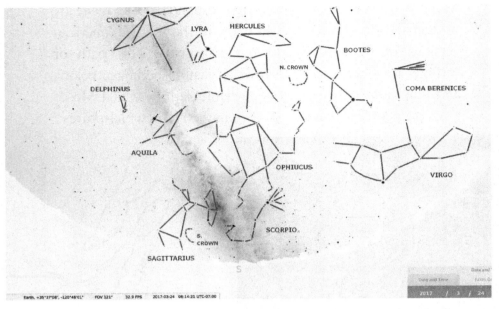

The frightening dog, which scares those trying to approach the underworld: Almost certainly Scorpio again, which may not seem to resemble a dog, but which probably does play the role of Cerberus in ancient Greek and Roman myth (see discussion in *Star Myths of the World, Volume Two*, in which this identification is made based primarily on descriptions from Hesiod and other early sources).

The kildeer, who also tries to startle travelers and make them fall from the rickety bridge into the river: Most likely Aquila, the Eagle, but possibly Cygnus the Swan. Both fly within the Milky Way, in the vicinity of the "rift path" identified above.

The rope of eagle-down with which the intrepid husband glides across the raging river, in some versions of the story: Probably the long "tail" of Aquila, possibly envisioned as connecting up to the western-half (the left side) of the serpent held by Ophiucus, to form a "rope" which stretches across the Milky Way by way of Aquila the Eagle.

The figure of Oscotarach, or "Pierce-Head," who draws out the brains of the dead when they arrive in the underworld: Probably Boötes, whose large bulbous head has a "pipe" or "spike" coming out of it -- which may be the origin of the figure of "Pierce-Head." In *Star Myths of the World, Volume Two* (which deals almost exclusively with the mythology of ancient Greece), extensive evidence was presented arguing that figures who are described as receiving a wound to the face, mouth, temple, or other part of the head in the Iliad (usually a fatal wound) can usually be definitively linked to Boötes (for instance, these characters are usually described as driving a chariot when they receive such a wound -- and Boötes is a constellation often envisioned as driving a chariot, with the horses being the bowl of the Big Dipper itself). Of course, it may be objected that the figure of Oscotarach is piercing the heads of *the dead* -- not having his own head pierced. Therefore, he himself may not be Boötes -- perhaps he is the constellation Hercules, who stands or crouches menacingly over Boötes while the latter is having his brain removed. On the other hand, Oscotarach and other "old man" figures encountered in the underworld journey as described in the Native American myths are usually *sitting* in their hut or next to a fire when they are encountered, which sounds like Boötes. In any case, "Pierce-Head" is probably associated with Boötes, if not actually identified himself with Boötes (if not Boötes, then he is at least nearby).

The calabash or gourd or pumpkin or box into which the soul of the deceased beloved can be stored for the journey back from the underworld: Most likely the Northern Crown. Other possibilities include the "head" of the snake held by Ophiucus (on the right-side of the figure of Ophiucus, as we face the star-chart above), as well as the large sword or club held by the figure of Hercules (remember that in the story, it is usually the "old man" figure, sometimes identified as Oscotarach the head-piercer, who gives the would-be rescuer the container to carry back the soul of the beloved deceased). Another possibility is the part of the constellation Sagittarius which is usually envisioned as the figure's bow. Since we have already seen that Sagittarius probably plays the role of the unsuccessful rescuer in some versions of the myth (because Sagittarius is "looking back"), it is possible that the soul-container, in some versions of the myth, is identified with the "bow" of Sagittarius.

The dead dancing in circles: The constellation Hercules, when envisioned as a "whirlwind." Previous books, especially the volumes of *Star Myths of the World, and how to interpret them*, have shown conclusively that the constellation Hercules was sometimes envisioned as a powerful figure bearing a great club, sword or mace, but other times as a whirling wind or a whirlpool or even a spider. This is because the square "head" of Hercules can be envisioned as having four whirling "arms" spinning out from a central hub, almost like a great pinwheel in the sky. Note also that the old woman figure in some versions of the story is named "Wind." See star diagrams on opposite page.

The skeletal wife covered with worms or maggots, or the skeletal figures of the dead in the underworld: Scorpio again – a constellation which often plays a worm, or a mass of worms, or a skeleton.

The buffalo kidney that the husband must give to the wife in order to restore her, according to the Comanche version of the story: The "head" atop the serpent of Ophiucus, which the figure of Hercules can be envisioned as "offering" to his wife (who may be either Ophiucus or Virgo, in this case). The head-end of the serpent of Ophiucus has been seen to play the role of a "human heart" in the Bible story of Solomon, as well as in the Sanskrit epic of the Mahabharata. This same feature probably also plays the role of the "tempting strawberry" in some North American versions of the oicotype. In fact, this celestial feature is probably also the pomegranate which Persephone encounters, and from which Persephone eats a few seeds, which causes her to have to remain in the realm of Hades for a certain number of months each year, as discussed in *Star Myths of the World, Volume Two*.

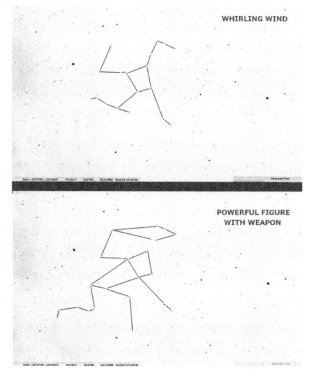

Above: two identical views of the same stars (constellation Hercules), connected with different imagined outlines, first as a spinning wind or whirlwind in top frame, and then as powerful warlike figure below.

The comb or combs which the husband throws down from his hair in order to escape from pursuit. The constellation Corona Australis (the Southern Crown), just behind the "fleeing" constellation of Sagittarius. We saw this detail in the version of the oicotype found in the Kojiki, in which combs are described as being from one side of his hair-knot and the other -- which probably references the "top-knot" or "hair bunch" seen atop the head of Sagittarius in the outline presented above (in the sky, it appears that the fleeing Izanagi has thrown down one comb, and still has the other in his hair). Note that combs appear in some versions of this same myth among the Native Americans.

From the above discussion, we can agree that it is ridiculous to assert that the Native American versions of the underworld retrieval myth were imported by early missionaries: the Native American versions clearly preserve far *more* celestial references than are present in the Orpheus-Eurydice version, for example.

It is also notable that, while the Orpheus-Eurydice version and the Izanagi-Izanami version both contain a taboo (or tabu) against looking at the deceased beloved, the various Native American versions contain a variety of different prohibitions: some of them prohibit touching, embracing, kissing or having sexual intercourse with the beloved within a certain period of time after the underworld escape, while others prohibit striking with the hand.

In many cases, these prohibitions can be traced directly back to a specific constellation -- but different constellations for different prohibitions! While the tabu against looking at (or looking back upon) the beloved is almost certainly associated with the outline of the constellation Sagittarius, the prohibition against striking with the hand is probably associated with the outline of the constellation Hercules.

Recall that in the Comanche version of the myth, the husband wishes to pull a buffalo blanket over himself and his beloved, and his hand slips and accidentally strikes his wife, causing him to lose her again. The act of pulling a blanket over the head is associated with the outline of the constellation Hercules, as I argue in *Star Myths, Volume Two*. In the Odyssey, the character Odysseus on more than one occasion pulls his cloak over his own head in order to weep -- and he is a figure who is very clearly associated with the figure of Hercules in the sky, as I show in that volume. When the husband in the Comanche story pulls the buffalo robe or blanket over his head, the great sword of Hercules becomes the blanket he is pulling over his head (and that of his wife, who is probably Ophiucus).

Note that the "lower hand" of Hercules reaches down towards Ophiucus -- this is almost certainly the hand that slips and strikes the wife. Hercules can be seen as "striking" Ophiucus.

It is interesting to ponder whether the admonition of the risen Lord Jesus in the New Testament, that Mary should not touch him until after he has returned to his Father, is an echo or remnant of this same ancient myth-pattern, in which the one who has been brought back from the dead may not be touched (see John 21: 17).

It should also be fairly obvious, as I in fact argue in *Star Myths of the Bible*, that the story of Lot's wife (who is admonished *not to look back* when she and her family are leaving Sodom, in Genesis 19) is another version of this same mythical oicotype. Once again, I believe we are dealing with a myth based upon

Sagittarius, a constellation whose outline makes it appear to be looking back over its shoulder. Sagittarius is Lot's wife.

The fact that the stories of the Bible (both in the Old and New Testaments, so-called) can be seen to be based on the very same celestial foundation which informs the various "retrieval from the land of dead" stories among the indigenous nations of the New World shows how misguided it is for Christian missionaries to dismiss them as "stupid stories" (as they are condescendingly termed in the translation of the writings of Jean de Brébeuf). The Bible stories have the same foundations!

And there are yet other myths from still other cultures which exhibit clear signs of belonging to this same oicotype. For instance, the story of the death of Baldr (or Balder) in Norse mythology, and especially the unsuccessful attempt by the god Hermod the Bold, son of Odin, to retrieve Baldr from the gloomy realm of the goddess Hel, contains many elements of the pattern.

In the account given by Snorri Sturluson in the prose Edda, we are told of Hermod "that he rode for nine nights through valleys dark and deep so that he saw nothing until he came to the river Gioll and rode on to Gioll bridge. It is covered with glowing gold."[74]

This bridge over the Gioll (an Anglicized spelling – it is more accurately rendered *Gjöl*, the noisy river that flows past the gates of the realm of the dead) is known as the Gjallarbrú, and it is guarded by the maiden Moðguðr. She challenges Hermod, saying that the bridge resounds so loudly under his feet that she knows he is not dead but living – and in addition to that his countenance does not have the coloration of the dead.

Hermod tells her that he has come to seek Baldr, and she tells him that he is going in the right direction. He proceeds to the gates of the underworld itself, spurs his horse to leap clear over the gate, and proceeds to the hall of the goddess of the underworld, where he finds her and nearby his brother Baldr, sitting in a seat of honor in the hall.

In this version of the myth, Hermod begs for the return of Baldr, explaining that all the gods are in great mourning at his loss. The goddess Hel declares that Baldr can indeed return, but only if all things in the world, alive or dead, show their sorrow by weeping -- but that if any object or any living soul should refuse to weep Baldr out of the realm of Hel, then Baldr must remain.

Upon Hermod's return with this message, the gods send messengers over the whole world, Snorri says . . .

> to request that Baldr be wept out of Hel. And all did this, the people and animals and the earth and the stones and trees and every metal, just as you will have seen that these things weep when they come out of frost and into heat. When the envoys were traveling back having well fulfilled their errand, they found in a certain cave a giantess sitting. She said her name was Thanks. They bade her weep Baldr out of Hel. She said:
>
>> "Thanks will weep dry tears for Baldr's burial. No good got I from the old one's son either dead or alive. Let Hel hold what she has."[75]

Thus Baldr had to remain in the underworld. It was determined that the pitiless crone must have been Loki in disguise, and the gods trapped him and bound him and inflicted a punishment on him similar to that inflicted upon Prometheus in the mythology of ancient Greece.

Here again we see a manifestation of the same failed retrieval, complete with a river that flows past the gates of the realm of the dead and that has a bridge that must be crossed -- undoubtedly the Milky Way flowing past the "doorway" framed by the stars of Ophiucus. We also see permission given to bring back the beloved deceased from the underworld, along with strict conditions that must be met -- and which are not met.

In the Baldr myth, we also see echoes of the pattern of the god who is slain (similar to the Osiris cycle of ancient Egypt), as well as the requirement that all the world must *weep* for the slain god (a pattern that is found in many other myths involving a slain god, including Tammuz or Dumuzi of ancient Mesopotamia; the weeping for Tammuz is mentioned in the Hebrew Scriptures, in Ezekiel 8: 14).

But, having established that this ancient myth-pattern is extremely widespread around the globe, it still remains to ask ourselves what knowledge or gnosis this myth could possibly be intended to convey to us?

A myth of such obvious importance probably contains many layers of profound meaning. Certainly commentators down through the years have noticed the incredible prevalence of this particular myth-pattern, and have offered various interpretations. Anna Gayton believes that "the story of a Visit to the Afterworld answers the question: what is the fate of the dead, as well as the skeptic's request to know how man came by this knowledge" -- and also reinforces the understanding that "there can be no intercourse between the two worlds."[76]

Åke Hultkrantz suggests that the stories may originally derive from shamanic healing rituals, in which the shaman travels to another realm in order to restore the sick -- and that perhaps "the rite for the curing of the sick has merged into a rite for the return of the dead."[77]

However, as practical as these suggestions appear, neither of them appears to have been formulated in light of any conscious awareness of the fact that this far-ranging oicotype can be shown to have undeniable *celestial* foundations. Without this understanding, I would argue that one of the very most important aspects of this myth-pattern will be overlooked. Having shown that the failed retrieval story consists of celestial metaphor, I believe we can then and only then explore possible interpretations in light of that understanding.

Perhaps the most striking aspect of each manifestation of the failed retrieval motif is the prescription of strict injunctions which must be followed by the visitor from the land of the living. In each case, the would-be rescuer attains the permission from an authority figure in the land of the shades to take back the departed beloved, but the permission includes a severe warning that must not be violated or the permission will be rescinded.

As we have seen in discussions found in the previous chapters, I am convinced that one layer of the message contained in the ancient Star Myths given to humanity concerns the understanding that this apparently-material universe in which we find ourselves is in fact intertwined with and interpenetrated by an invisible realm, an infinite realm, at every point and at every level.

In the language of quantum physics (as discussed in my 2014 book, *The Undying Stars*), this invisible or infinite realm can be thought of as a realm of *pure potential* (or "pure potentiality"), a realm of infinite possibility but also of non-manifestation.

According to the understanding of quantum physics that has been developed over the past hundred years in response to the results of thousands of actual experiments, there seems to be a state of "superposition" in which particles can exist in a condition of "potentiality," out of which state or condition they can then "manifest."

Metaphorically, some ancient cultures refer to the invisible realm as the "seed realm," which is a brilliant way of describing this concept. When a tree or a plant is still in seed form, how many branches or leaves does it have? The answer is that we don't really know: the tree or plant might have any number of branches or leaves once it grows to maturity. But when it is in its seed form, the number of possibilities is practically endless. When it is in seed form, it is still in the realm of "potentiality." Once it grows into a plant or a tree, it will have a defined, concrete number of branches or leaves -- it will have left the realm of "potential" and manifested in a specific form, in the material world.

Similarly, quantum physics arose from experiments which showed evidence of subatomic particles behaving in much the same way -- as if capable of existing in a state of "superposition" or "pure potentiality," capable of manifesting in one form or another, or in one or another of two "box pairs." Once they manifest in one state or another, they exit the realm of pure potential -- but until they manifest, they seem to exist in a state of potential in which more than one outcome is possible.

This understanding helps us to grasp the relationship depicted in the myths (including in the "failed retrieval" myths) between the immaterial realm and the material realm. If we think back to the discussion in the preceding paragraphs about the "seed realm" or the realm of "pure potentiality," we can readily admit that the manifestation of the tree or the plant can be said to proceed from the seed realm (the realm of potential). The potential comes first: the manifestation flows from the realm of pure potentiality. One is dependent upon the other -- and it is the material or manifest state that "depends upon" or "flows from" the immaterial or non-manifest state.

We have abundant testimony from the ancient wisdom contained in the myths -- and from representatives of traditional cultures still connected to the ancient wisdom of their ancestors -- that this relationship was well understood in cultures where the ancient wisdom was preserved. In the record of the teachings of the Lakota holy man Black Elk, for example, the relationship between this material realm and the invisible and infinite realm is very clearly stated. Describing the vision of his illustrious relative, the warrior and leader Crazy Horse (Tashunke Witkó) of the Oglala, Black Elk says:

> Crazy Horse's father was my father's cousin, and there were no chiefs in our family before Crazy Horse; but there were holy men; and he became a chief because of the power he got in a vision when he was a boy. When I was a man, my father told me something about that vision. Of course he did not know all of it; but he said that Crazy Horse dreamed and went into the world where there is nothing but the spirits of all things. That is the real world

that is behind this one, and everything we see here is something like a shadow from that world. He was on his horse in that world, and the horse and himself on it and the trees and the grass and the stones and everything were made of spirit, and nothing was hard, and everything seemed to float. His horse was standing still there, and yet it danced around like a horse made only of shadow, and that is how he got his name, which does not mean that his horse was crazy or wild, but that in his vision it danced around in that queer way.[78]

This passage is remarkable, and worthy of deep contemplation.

For one thing, it is describes the vision by which the great leader Crazy Horse received his name.

Further, it provides tremendous insight into the very sophisticated understanding of the "world of spirit" or of shadow. Of this world, Black Elk declares: "That is the real world that is behind this one, and everything we see here is something like a shadow from that world." In other words, this material world is actually a *derivative* world -- it takes its shape from the spirit world, it flows from that world, and it is only a shadow of that *other* world, which is in fact the real one.

The material realm *depends upon*, and *proceeds from*, the world where there is nothing but the spirits of all things. In a sense, the material realm *takes its orders from* the spirit realm.

We see this relationship depicted very clearly in the world's ancient myths. The material realm depends upon the realm of spirit -- the realm of the gods -- and the order between the two must never be inverted.

In the Greek myths, for example, we can all think of countless examples of stories in which a human declares that their skill or beauty or musical ability is greater than that of a god or goddess -- with disastrous results. Such a declaration is an affront to the realm of the gods, because the beauty or the skill or the talent *came from* the realm of the gods in the first place. Thus, to say that a mortal maiden is more beautiful than the goddess of beauty, or to proclaim that a mortal musician is more skillful than the god of music, is to elevate the recipient of the gift above the source of the gift. It is an inversion of the proper order.

The Star Myths of the world use the infinite heavens as a beautiful and awe-inspiring *visible* representation of the *invisible* realm within which this visible realm has its *source and origin*. It is the realm of pure potential, from which everything that manifests in the material realm proceeds. The lower realm has its pattern in the higher realm – the realm of manifestation has its origin in the realm of pure potentiality.

For this reason, the Hermetic tradition proclaims: "As above, so below." The saying teaches us that the material realm takes its pattern from the invisible realm. The ancient wisdom never says, "As below, so above." It is the infinite realm which is the source for that which is manifest: the infinite gives the pattern.

And this very same teaching, I believe, is being conveyed in these various representations of the "failed retrieval" pattern. It is the realm of shadows, the realm where nothing is material, which dictates the conditions to the representative from the material realm. The visitor from the physical world does not dictate to the representatives of the world of shades.

And this myth of the failed retrieval conveys so much more, if we begin to follow the metaphor implied by the celestial content yet further.

We have seen in our examination of the celestial correspondences themselves that in some versions of this oicotype, the would-be rescuer (such as Orpheus) *looks back* just as the difficult journey to the underworld and back has almost reached its successful conclusion, and that this looking back motif is very characteristic of *the outline of Sagittarius*, a constellation which indeed seems to look back over its own shoulder.

Significantly, we have already seen that on the great wheel of the zodiac as oriented during previous ages, Sagittarius was a constellation associated with the "great turning point" of the winter solstice: that point where the downward progress of the sun finally reaches its limit, and a great pivot in the cycle sees the downward motion reverse and begin the upward journey once again.

Sagittarius, in fact, is located just before the point of winter solstice (in the Age of Aries) -- the point of new birth. Additionally, Sagittarius is the first zodiac constellation located on the east side of the great widening of the Milky Way band which is indicative of the Galactic Core, (and the east side of the Great Rift, which was seen as a celestial birth canal in some ancient cultures, including perhaps the Maya).

Thus, Sagittarius is representative of the "new birth" and of the "turn back towards spirit" after the plunge into matter that commences (metaphorically speaking) at the fall equinox point presided over by Virgo. Sagittarius is thus located at or near the *darkest point* of the "lower journey":

DAYS LONGER THAN NIGHTS:
Heaven, Promised Land, Greece, etc.

NIGHTS LONGER THAN DAYS:
Hell, Egypt, Troy, etc.   SAGITTARIUS

The sun's progress through the year is indicated by the two small arrows on the outer edge of the circle in the preceding diagram: the sun moves clockwise through the zodiac chart as it is depicted in the illustration on the previous page. That means that the "underworld journey" which begins at Virgo and the "plunge" of the autumnal equinox then proceeds in a clockwise fashion down through Sagittarius, before turning back "upwards" towards the opposite equinox between Pisces and Aries.

Thus, Sagittarius can be seen as a point of "maximum doubt," where the need to proceed ahead and *not look back* is the greatest. The great turning point, the great pivot of the year, has been reached – emblematic of the recognition of the spiritual nature, and the decision to begin to elevate the divine aspect as opposed to continuing to act as though the material is all only aspect of our nature. At this point of "turning back upwards," the doubt expressed by looking back longingly (as Lot's wife looks back upon Sodom) is especially dangerous.

I am convinced that the dramatic action in these powerful, poignant, and painful "underworld retrieval" myths is intended to caution us against "looking back" once we have begun the upward path. These myths are a warning against becoming too attached to the lower realm.

The upward journey of Orpheus and Eurydice represents the elevation of both spirit and body – an important aspect of our own "descent into the underworld" which we each undergo during our plunge into this incarnate life.

This upward elevation of spirit and body can be accomplished during our earthly sojourn -- but to do so successfully, we cannot give in to doubt, or to fear or to inordinate desire and emotion.

And, in order to successfully do so, we must follow the pattern given to us *from the realm of the gods*, taking our cue from the invisible realm – and (as we will now see) from our *Higher Self*.

The Divine Twin

As we begin to delve deeper into the subject of how the ancient wisdom encoded in the myths may apply to our own lives, in ways that are at once both practical and profound, we begin to see that this entire metaphorical system -- in which the infinite heavenly realm of the stars becomes a visible representation of the Invisible Realm of "pure potentiality" (the realm of the gods) -- points us towards an awareness that this universe or cosmos in which we find ourselves is intertwined and interpenetrated at every point by that Infinite and Invisible Realm . . . and that *we ourselves* must therefore have an invisible and infinite component as well.

Indeed, it would be a grievous omission if we were to examine all these myths from around the world, seeing all of their incredible connections to the stars and all of their astonishing parallels with one another, and fail to perceive that one of the central themes running through the ancient wisdom imparted to cultures on every inhabited continent and island of our globe involves the integration of our ordinary, uncontroversial, mortal and physical selves with an invisible, infinite, immortal and spiritual nature -- a nature which might be called the "higher self" (and which in some cultures is explicitly given a name which means exactly that).

While the concept of the higher self has now been talked about enough in some circles to risk becoming cliché, especially if it is not explained or understood in a way that enables us to actually apply its powerful meaning to our own lives, the ancient myths provide us with a plethora of illustrations and metaphors that can enable us to grasp the teaching at a very deep and practical level.

Any student of myth who has examined a wide range of sacred stories from many cultures will soon realize that there is a "superabundance" of myths involving "twinning" -- tales in which actual twins, or very close and intimate pairs who may not have been born twins but who function very much like twins, feature prominently or play a central role. The list of such twins in myth is almost endless, and would include the following (as well as many more):

- ➢ Gilgamesh and Enkidu in ancient Mesopotamia
- ➢ Jacob and Esau in the Hebrew scriptures
- ➢ the Ashvins in the ancient Vedas
- ➢ Castor and Polydeuces in ancient Greece
- ➢ Achilles and Patroclus in the Iliad
- ➢ Jesus and Thomas Didymus in the New Testament
- ➢ Hunahpu and Xbalanque in the Popol Vuh of the Maya
- ➢ Taiwo and Kehinde of the Yoruba
- ➢ Romulus and Remus of ancient Rome
- ➢ Amphion and Zethus of the ancient cult of Thebes
- ➢ David and Jonathan in the Hebrew scriptures

Very frequently, one of the two twins will be divine (or at least semi-divine) and the other will be mortal. Sometimes, one will be described as having a divine parent, while the other will have two mortal parents.

For example, the hero Heracles (or Hercules) is a son of a divine father, Zeus, and a mortal mother, Alcmene (or Alcmena). Heracles in ancient myth also has a twin-brother, Iphicles, who is not semi-divine, being the son of a mortal father, Amphitryon (who is the foster father of Heracles as well), and Alcmene.

Similarly, Kastor and Polydeuces (or Castor and Pollux, to use the Latin form), are described in some versions of the myth as being mortal (Castor) and immortal (Polydeuces). This is because, like Heracles and Iphicles, Polydeuces is the son of the divine Zeus and the beautiful mortal woman Leda, while Castor is the son of Leda and her mortal husband Tyndareus, king of Sparta.

Note that Leda is famously seduced by the god when Zeus takes on the form of a swan -- and the constellation Cygnus is located immediately adjacent to the constellation Hercules (who plays the role of Zeus in most of the Greek myths, as discussed in Star Myths of the World, Volume Two). Cygnus is a constellation which appears to be flying "downwards" in the Milky Way, towards Sagittarius and Scorpio; Leda probably corresponds to Sagittarius in the story of Leda and the Swan.

In the most famous part of the myths of Castor and Pollux (Poldeuces), the mortal brother, Castor, is slain in a battle, and Polydeuces appeals to Zeus to allow Polydeuces to share some of his own divine nature with Castor, in order to save him. Here is how the ancient poet Pindar (c. 517 BC - c. 437 BC) describes the arrangement, from the tenth Nemean Ode, as translated by Diane Arnson Svarlien:

> Swiftly Polydeuces the son of Tyndareus went back to his mighty brother, and found him not yet dead, but shuddering with gasps of breath. Shedding warm tears amid groans, he spoke aloud: "Father, son of Cronus, what release will there be from sorrows? Order me to die too, along with him, lord. A man's honor is gone when he is deprived of friends; but few mortals are trustworthy in times of toil to share the hardship." So he spoke. And Zeus came face to face with him, and said these words: "You are my son. But Castor was begotten after your conception by the hero -- your mother's husband -- who came to her and

sowed his mortal seed. But nevertheless I grant you your choice in this. If you wish to escape death and hated old age, and to dwell in Olympus yourself with me and with Athena and Ares of the dark spear, you can have this lot. But if you strive to save your brother, and intend to share everything equally with him, then you may breathe for half the time below the earth, and for half the time in the golden homes of heaven." When Zeus had spoken thus, Polydeuces did not have a second thought. He opened the eye, and then released the voice of the bronze-clad warrior, Castor.[79]

Here we encounter a strikingly powerful image: the divine twin who rescues the mortal twin, sharing his immortality with his mortal counterpart. I am convinced that this pattern is given to us as an illustration of the reality of our condition while passing through this incarnate life. We are encased in a mortal form, but we have access to a "divine twin" who is ready to lift us up and allow us to share in that same divine nature. In fact, in the story of Castor and Polydeuces, Castor also becomes a divine figure -- and the Twins of Gemini are of course visible in the night sky in the zodiac constellation whose two brightest stars bear the names of Castor (the dimmer of the two) and Pollux (the brighter).

My understanding of these sacred stories indicates that the two natures are dramatized in the myths as two distinct individuals, but that in fact the truth being conveyed is that they correspond to the two natures inherent in *one person* -- the two natures which every single man and woman himself or herself possesses, a "lower self" and a "higher self," or a "mortal self" and a "divine self."

This concept is dramatically illustrated by the New Testament writer who calls himself *Paul*, and who is described in the narrative contained in the book of Acts as having previously been called *Saul*. As Saul, he was antagonistic to Jesus, but as Paul he was integrated with and empowered by the "Christ within." His previous alienation from his divine nature is dramatized by the encounter in which Jesus appears to Saul on the road to Damascus, and asks Saul: "Why do you persecute me?" -- and Saul is transformed.

The two aspects of Paul are almost like two different people -- two "twins" named *Saul* and *Paul*. But they are really the same individual, and the story is intended to illustrate a profound truth for our understanding.

As the brilliant and insightful Robert Taylor notes, in a sermon he delivered on January 9, 1831 and which is recorded in *The Devil's Pulpit*, the names of Saul and Paul have undeniable celestial antecedents. The name *Saul*, of course, clearly recalls the Saul in the Old Testament, who persecuted David (David and Jonathan being another manifestation of "the Gemini or Twins" in the story, according to Taylor) -- but the name also recalls the syllable *sol*, which corresponds to the sun.[80]

The name *Paul*, on the other hand, corresponds to both Apollo (the sun god of ancient Greece and Rome) and to Pollux or Polydeuces!

As Taylor elaborates:

> And Saul and Paul are one and the same persons, only as the Sun of November is the same as the sun of May. Only in different characters: Saul before his conversion, being the November sun, in the sign of Sagittarius, where you see the Great Persecutor, with his bow and arrow, playing havoc with vegetable nature, stripping the trees of their foliage, riding down to Damascus, and on the high road to

hell and Tommy -- that is, St. Thomas's day, which is the 21st of December, the lowest point of the sun's declension: and, consequently, the lowest pit of hell.

The same Saul being, in Hebrew, the self-same word

שאול

which is, wherever it serves the purpose, translated hell [*Sheol*, which is translated in the King James version sometimes as "hell," sometimes as "pit," and sometimes as "grave"]: as the Greek name Paul is an abbreviation of the Greek Apollo, under whose protection the month of May is placed in the calendar of Julius Caesar, and of the name of the star Pollux, in which the sun appears in his regenerate and mild and amiable character at that delightful season.[81]

Taylor's analysis reveals that the "lower self" (in this case, Saul before his conversion, alienated from and even "persecuting" the divine nature) corresponds to the lowest point on the annual cycle -- in which the sun is cast down into the sign of Sagittarius just as Saul in the book of Acts is cast down and blinded when he is on the Damascus Road -- while the "higher self" (the transformed Paul) corresponds to the sign opposite Sagittarius on the zodiac wheel, Gemini, through which the sun travels in May and June, on its way to the highest point on the annual cycle.

As discussed in *The Undying Stars* (2014), opposite signs on the zodiac are sometimes described in myth as "adversaries" or opponents, and in this case the metaphor is especially apt, as the sun is "persecuted" in Sagittarius (just prior to the lowest point on its cycle) and "exalted" in Gemini (just prior to the highest).[82]

And yet, as Taylor points out, it is expressly stated in Acts 13: 9 that Saul and Paul are one and the same person. The two together are intended to illuminate something about our own human condition. As we have seen in the extended quotation from Alvin Boyd Kuhn regarding the symbology of the annual cycle, the lowest point (where the sun is cast down in Sagittarius, as Saul is cast down on the road) is the great "turning point," representative of the point in our lives at which we begin to become aware of the existence of the higher nature and the importance of the Invisible Realm -- the point where the Djed column begins to be re-established after being cast down, the point at which we begin to elevate and integrate our divine nature once again, after a period of temporary alienation and disassociation.

As the quotation above from Taylor's 1831 sermon points out, the 21st of December was traditionally observed as the day of St. Thomas (and remains on that day in some Anglican churches) -- indicating an understanding at some level that Thomas in the New Testament is also associated with the "lower self," dramatized (I believe) in the well-known episode of "doubting Thomas" in the gospel accounts.

This episode of Doubting Thomas is discussed at length in *Star Myths of the World, Volume Three* (*Star Myths of the Bible*).[83] The important insight into the significance of the identity of Thomas as it pertains to the discussion in this chapter comes from the observation that Thomas in the gospel texts is explicitly said to also be "called Didymus" -- which means "the Twin" (the prefix *di-* meaning "two"). From this, we can see that we are probably dealing with the same theme involved in the other twin stories from the world's mythologies.

The texts that were selected to be part of the New Testament "canon," however, give no indication as to the identity of the twin of Thomas. The illuminating answer is found in an ancient text which was excluded from the canon by the authorities of the church hierarchy during the fourth century, but which survived in a large sealed jar buried beneath the sands at the base of a cliff near the modern-day village of Nag Hammadi in Egypt (no doubt to escape the persecution of those same church authorities, who outlawed texts that had not been selected), and rediscovered in the twentieth century.

In the Book of Thomas the Contender (a title which itself recalls the above discussion of the "adversary" or "persecutor"), Jesus addresses Thomas and tells him directly that Thomas is "my twin and true companion."[84]

This astonishing understanding, that Thomas is in some way the "twin" of Jesus, opens up an entirely new perspective on the meaning of the confrontation between Thomas and the risen Christ described in the gospel account of John chapter 20.[85]

In that encounter, Thomas is filled with doubt, and declares to the other disciples: "Except I shall see in his hands the print of the nails, and put my finger into the print of the nails, and thrust my hand into his side, I will not believe." But Jesus appears and directs Thomas to "Reach hither thy finger" and feel the wounds, and to "reach hither thy hand" and thrust it into his side -- and Thomas answers with the declaration, "My Lord and my God" (John 20: 25 - 28).

Drawing on what we have already observed in this chapter's discussion of the Twins of Gemini and of Saul and Paul, we can deduce that if Thomas is specifically said to be a twin, and if Jesus himself tells Thomas that Thomas is the twin of Jesus, then we can conclude that we are again dealing with a situation with a divine twin (Jesus) and a mortal twin (Thomas) – and that the two are illustrating a truth about our own situation in this incarnate life. Like Saul and Paul, they are not illustrative of two different people, but of the "lower self" (Thomas) who rejects the higher self (Jesus, or the "Christ within") and who later becomes reconciled with and guided by the higher self, the divine nature.

As illustrated in *Star Myths of the Bible*, artists down through the centuries have depicted this encounter between Thomas and Jesus in ways that indicate an identification of Thomas with the outline of the constellation Capricorn -- the zodiac sign which actually commences at the December 21 juncture between Sagittarius and Capricorn (traditionally Thomas's day). This fact only reinforces what we have already seen about the symbology of the "upper and lower halves" of the zodiac wheel, and the significance of the great "turning point" at the winter solstice where the Djed column begins to be restored, and our connection with the higher nature has its inception.

In the illustrations on the opposite page, we see painting by Giovanni del Giglio from the early 1500s of the *Incredulità di San Tommaso* ("Incredulity [or disbelief] of St. Thomas"), in which the outstretched fingers of the hand of Thomas and the downturned fingers of the hand of Jesus mimic the outline of the Goat of Capricorn. In fact, the outline of Capricorn in the sky appears as two triangles, and this painting contains both of them, along with the horns of the Goat and its upturned tail (the tail indicated by the hand shape of the woman behind them).

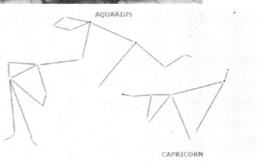

Note also the fact that the horns of Capricorn point directly towards the "forward leg" of Aquarius -- which can be seen as a spear going directly into the side of that constellation. Thomas, of course, is directed to put his hand in the side of Jesus, where the spear went in, reminiscent of the relative positions of Capricorn and Aquarius.

The understanding that the "higher twin" and the "lower twin" are in reality depicting the experience of one person is reinforced by a widespread cultural detail reported in a valuable study of the mythological significance of twins and of the myth-pattern of the divine twins typified by Castor and Polydeuces, published in 1906 and written by James Rendel Harris, entitled *The Cult of the Heavenly Twins*.

In that book, the author notes that in many traditional cultures around the globe, twins in previous generations would *both be given the same name!*[86] Harris devotes an entire chapter to the naming of twins in different traditions around the world. In some cultures, the same name would be given to both (and he cites passages from Saxo Grammaticus, Bede, and the Heimskringla saga to illustrate this practice), whereas in others the same two names would be given to all twins, one name always given to the firstborn and the other name always given to the second of the pair.[87]

The study by Harris also notes instances in which mythological twins are described or depicted as being of greatly differing physical ages, especially as they reach adulthood. For instance, he cited examples in which the Dioscuri (a title given to Castor and Polydeuces, meaning "sons of Dios" or "sons of Zeus") are depicted by ancient artists as being of very different ages, with one having a beard and appearing much older than the other who does not. Harris even cites a description by the ancient geographer Pausanias (c. AD 110 - c. AD 180) attesting to such a depiction of the Dioscuri on the Chest of Cypselus, a richly decorated chest allegedly offered to the gods by Cypselus of Corinth, who died in 627 BC, and whose name itself means "chest" or "corn-bin" (allegedly he was hidden in the chest as an infant, to escape harm during a palace coup).[88]

What could be the significance of such a discrepancy? Harris also notes that in some cultures, there is "a curious and impossible tradition about the birth of twins, where the interval between the elder and the younger is imagined to be as long as three months" -- and he also notes parallels in the Talmudic literature, where the second twin at one point is said to have been born thirty-three days after the first![89]

Following the analysis of Alvin Boyd Kuhn, presented in *Lost Light*, it is very likely that the traditions in which the twins are impossibly said to have their births separated by a symbolically significant period of time (such as thirty-three days, or even three months), which is literally impossible but metaphorically meaningful, dramatizes in fable the understanding that the "twin natures" of each man or woman have their "birth" at different points during life.[90]

The birth of the mortal, physical person comes first, of course: when the spirit plunges down to incarnate in a body of flesh (symbolically associated with the point of fall or autumn equinox in the metaphor underlying the ancient myths). The birth of the spiritual awareness takes place later -- at the point of "second birth" -- metaphorically represented by the point of winter solstice, the great pivot-point on the zodiac wheel. These points are in fact three months apart in the twelve-month cycle!

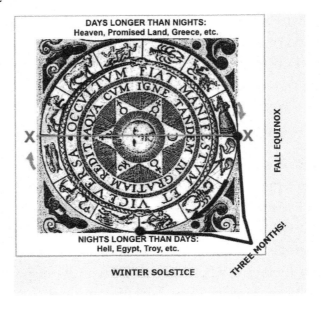

Based on this understanding, that the two twins actually represent two "births" and two "natures" within one individual, we can begin to understand the prevalence in mythology and in folklore of the pre-eminence of the child born second or last.

For example, in the story of Jacob and Esau (who are twins), Esau is born first but Jacob is given the blessing as if he is the elder. Additionally, Esau is described as hairy and physically powerful and vigorous, while Jacob is described as "smooth" and less interested in feats of physical prowess. Harris points out that these descriptions harmonize with the depiction of twins such as the Dioscuri or the Kabiri (or Kabeiroi, the "Great Gods" whose mysteria were observed at Samothrace, and who share many characteristics with the Dioscuri) as being of different apparent ages, with one bearded and the other "smooth."[91] He notes the same difference between the Theban twins Amphion and Zethus.[92]

In these stories, it is the younger who is treated as special – not because younger siblings are more special than older siblings, but because these myths contain and convey an esoteric truth, that the "second birth" (in which we connect with our divine nature and our higher self) is the focus of the teaching, and the birth we should be seeking, rather than our physical birth (by which we obtain a physical body, symbolized by the "hairy twin" in the story, such as Esau or Zethus or Enkidu).

Along these same lines, Alvin Boyd Kuhn points to the story of the Prodigal Son in the parables of Jesus recorded in the New Testament gospels. Here again it is the younger brother who is embraced and exalted by the father – creating resentment in the elder brother similar to the case with Esau's jealousy of Jacob.[93] In this story, Kuhn points out the additional twist that the exaltation comes only after the prodigal younger son goes off and lives like an animal, indulging in wanton sexual behavior and ending up eating with pigs at the trough.

The meaning of this parable is not easy to decipher, Kuhn notes, because the elder brother has been obedient and yet it is the younger who is given the reward and the feast. The solution, he argues, is to understand that in this parable the younger brother represents the soul who has descended into the material world, and who then reaches the "turning point" and wakes up from the stupor of animal behavior, enabling a change in the spirit which could never be achieved without experiencing incarnation![94]

Said another way, the ancient myths seem to teach that it is the "descent" on the great wheel -- the "first three months" so to speak -- which enables a different level of spiritual growth and ascendance than could be achieved by remaining in the "upper half" of the circle and never experiencing incarnation.

The myths depict this truth in many variant forms, but the message always seems to be that one of our central missions in this incarnate life involves the elevation of the divine nature -- in ourselves and also in others -- and that one of the most important aspects of this mission is our connection with the divine twin, the higher self.

Even if we accept the centrality of such a teaching in the ancient myths, however, the question still remains regarding the actual process by which we can seek such elevation of the spiritual nature and integration with the higher self. It is all well and good to realize that like Castor we can share the immortality of Polydeuces, or that like the prodigal in the parable we can wake up from eating husks at the trough, but how exactly do we proceed from there?

We will take up that subject in the subsequent chapters. For this chapter, it is enough to establish beyond reasonable doubt the centrality of this vital theme in the ancient myths, often embodied in stories involving a pair of twins, but also depicted in the form of many other "divine-to-mortal" relationships, all of them ingeniously conceived for our benefit and blessing.

Eros and Psyche, Charles Paul Landon (1760 - 1826).

The Higher Self

If the ancient myths, scriptures, and sacred stories of the world teach the existence of a "divine twin," a *higher self* -- and the evidence overwhelmingly establishes that they do teach such a doctrine -- then how do we establish greater integration with this higher self? And does such a concept have any practical use for our daily life?

As it happens, we are not entirely on our own with this question -- the very same ancient wisdom which points us towards an awareness of the existence of the higher self also provides guidance for connecting with this divine twin, should we choose to avail ourselves of that guidance.

In the Katha Upanishad of ancient India, we read that a virtuous mortal youth named Nachiketa is sent to Yama, god of death, by his angry and impulsive father -- and waits in the house of Yama for three days without eating or drinking. When the god returns, impressed by the discipline Nachiketa has shown, Yama offers three boons to Nachiketa.

Nachiketa first asks that his father would be calm cheerful, and free from anger towards his child. He then asks Yama to explain all the aspects of the fire ceremony, which is the support of the universe, and the path to heaven. For the third and final request, Nachiketa asks about the nature of existence after death.

Yama tries to dissuade Nachiketa from asking about this subject, offering him wealth, power, long life, fair maidens -- but Nachiketa insists, and after seeing his resolve, Yama compliments him on his desire for knowledge and his refusal to be tempted away from it by the offer of pleasures and wealth, a road on which many others sink.⁹⁵

Yama then begins to instruct Nachiketa, and his instructions center on the subject of Atman:

> Many there are who do not even hear of Atman; though hearing of Him, many do not comprehend. Wonderful is the expounder and rare the hearer; rare indeed is the experiencer of Atman taught by an able preceptor.
>
> Atman, when taught by an inferior person, is not easily comprehended, because It is diversely regarded by disputants. But when It is taught by him who has become one with Atman, there can remain no more doubt about it. Atman is subtler than the subtlest and not to be known through argument.
>
> This Knowledge cannot be attained by reasoning. [...]
>
> Atman, smaller than the small, greater than the great, is hidden in the hearts of all living creatures. A man who is free from desires beholds the majesty of the Self through tranquility of the senses and the mind and becomes free from grief.
>
> Though sitting still, It travels far; though lying down, It goes everywhere. [...]
>
> Know the atman to be the master of the chariot; the body, chariot; the intellect, the charioteer; and the mind, the reins.
>
> The senses, they say, are the horses; the objects, the roads. The wise call the atman -- united with the body, the senses and the mind -- the enjoyer.
>
> If the buddhi, being related to a mind that is always distracted, loses its determinations, then the senses become uncontrolled, like the vicious horses of a charioteer.
>
> But if the buddhi, being related to a mind that is always restrained, possesses discrimination, then the senses come under control, like the good horses of a charioteer.
>
> If the buddhi, being related to a distracted mind, loses its discrimination and therefore always remains impure, then the embodied soul never attains the goal, but enters into the round of births.

> But if the buddhi, being related to a mind that is restrained, possesses discrimination and therefore always remains pure, then the embodied soul attains that goal from which he is not born again.[96]

This same passage is also quoted in *Light on Yoga*, by B. K. S. Iyengar (first published in 1966), in which next to the word Atman, which is capitalized, is written the word "(Self)" -- also capitalized.

Clearly, in the above passage from the Katha Upanishad (or Kathopanishad), Yama is teaching a distinction between "body, sense and mind" and something called "the enjoyer," which is somehow separate not just from body and senses (which is fairly easy to understand) but also from "mind" -- which is much less intuitive.

We don't have much difficulty making a distinction between our "Self" and our body or our senses. However, we are usually accustomed to thinking of "ourselves" as being identical to or co-equal with our "mind." For example, when we say we appreciate someone for their mind, we usually mean that we appreciate them not for their physical beauty or physical body, but rather for who they are *inside* -- who they *really* are.

And yet this passage is apparently making a distinction between our mind and our Self. Is our true Self somehow distinct not only from our body and our senses but also from our mind?

The distinction is even more evident in the English translation provided in the introduction to *Light on Yoga*:

> Know the Atman (Self) as the Lord in a chariot, reason as the charioteer and mind as the reins. The senses, they say, are the horses, and their objects of desire are the pastures. The Self, when united with the senses and the mind, the

wise call the Enjoyer (Bhoktr). The undiscriminating can never rein in his mind; his senses are like the vicious horses of a charioteer. The discriminating ever controls his mind; his senses are like disciplined horses.[97]

How can we understand this distinction, which teaches that we ourselves are not the same as our mind? Or, put another way, if we are accustomed to thinking of ourselves as co-existent with our "mind," then what definition of "mind" are the above ancient scriptures using, since they obviously are *not* using the word "mind" to mean our True Self?

Obviously, what these ancient writings are calling the "mind" is something from which the Self stands apart. The mind is compared to the reins which guide the horses of the senses -- but there must be something above the reins which actually guides the reins themselves, and thus something above the mind.

In another ancient Sanskrit text, the Mahabharata (which contains the beloved Bhagavad Gita), the charioteer who guides the chariot of the semi-divine Arjuna during the battle of Kurukshetra is Lord Krishna himself. It is Krishna who delivers the Bhagavad Gita -- the Song of the Lord -- to Arjuna in the pause before the beginning of that great battle, a battle which I am convinced is also metaphorical and represents this incarnate life, just like the other epic battles in ancient myth, such as the Trojan War described in the Iliad.

The divine Krishna and the hero Arjuna have a very close relationship in the Mahabharata, and in fact exhibit many of the characteristics of the "twinning" pattern discussed in the previous chapter -- with Krishna clearly the "divine twin" in the relationship.

Thus, combining our understanding of the concept of the divine twin with the teaching being delivered in the Katha Upanishad, we can deduce that the Self who is described in the quoted passage above as "the Lord in the chariot" is in fact the Higher Self, or (more accurately) the Self who is united to the Higher Self – functioning as an integrated pair or team, like Krishna and Arjuna. Because, as we saw in the discussion in the previous chapter, the twinned pairs in these myths actually represent two aspects of one individual – even closer than brothers, even closer than twins, they are in fact one person (as the verse in the Proverbs tells us: "there is a friend *that* sticketh closer than a brother").[98]

The mind, which we tend to think of as *who we are*, is clearly very important. We don't want to denigrate it, and we certainly don't want to get rid of it. In the metaphor provided in the Katha Upanishad, the mind is the reins -- and we certainly don't want to get rid of those, if we are guiding a chariot drawn by a team of powerful horses.

Apparently, it is not unusual for us to think of ourselves as co-equal with our mind: as Yama tells Nachiketa, many have not even heard of Atman, and of those who have, many do not comprehend.

Unaware of the existence of the Higher Self, many *identify* with the mind -- and I believe that the modern label, introduced by Freud, of "the ego" is a useful term for us to use in this discussion. The word *ego* itself comes from the first person conjugation of the Latin verb for *to be*, which in English would correspond to "I am."

This "ego-mind" (as we might call it) is portrayed in the ancient myths as being impulsive, as being generally well-meaning (though often misguided), and above all as being *doubtful*, filled with doubts and conflicted thoughts.

In the story of "doubting Thomas," for example, we have an extremely memorable depiction of the "lower self" or "ego-mind." Thomas is impulsive, headstrong, and most notably extremely skeptical of the claims of the other disciples that they have seen the risen Lord.

"Except I shall see in his hands the print of the nails, and put my finger into the print of the nails, and thrust my hand into his side, I will not believe," Thomas declares, as we saw in the previous chapter.

We should all be able to agree that this aspect of the mind is not entirely negative, and that in fact it serves an important purpose. We *want* our mind to be able to assess and weigh different possible courses of action, and to be able to skeptically assess claims made by others, and even to present to us doubts about what we ourselves are doing.

If we are driving away from the house to go on vacation, and we ask ourselves, "Did I forget to turn off the burner on the stove after I made those fried eggs this morning?" this doubt might actually be appropriate. Our doubt, in this case, might actually save us from burning down the neighborhood, if it turns out that we did in fact fail to turn off the burner and our doubt caused us to turn around and check it.

However, if these aspects of the ego-mind get out of control, they can be detrimental to our ability to function at our highest level, and even detrimental to our ability to accomplish things that we could accomplish in life.

The tendency of the mind to have doubts is positive if it keeps us from burning down the house -- but if we go back to check the stove three times, we would probably agree that something is out of balance. And, the doubts of the ego-mind are no help to us if we are trying to shoot a winning three-pointer in the last seconds of a basketball game. Our tendency to doubt ourselves, to be overly critical and overly pessimistic, and to *over*think, can also hold us back from opportunities, whether these involve talking to someone to whom we might be romantically attracted, or whether they involve pursuing a dream that might be tremendously rewarding and fulfilling (and one that might help many other people as well).

In the story of Jesus and Thomas Didymus, when Jesus confronts Thomas, he does not tell Thomas, "You bad disciple -- because you doubted you can never have any part of me any more." Instead, he dispels Thomas's doubts, and Thomas responds by declaring, "My Lord and my God."

In other words, Jesus restores Thomas to the correct relationship among the "Lord of the chariot," and the "charioteer of reason (or, the intellect)," and the "reins that are the mind." If the mind and the reason (or the intellect) are steered by the Higher Self -- the divine twin, the Lord in the chariot -- then the proper relationship is enjoyed, and the powerful horses of the senses become disciplined and beneficial, instead of vicious and out of control.

Another powerful portrayal of the mind's tendency to let its endless back-and-forth and its doubts get out of control is presented in the myth of Eros and Psyche (or Cupid and Psyche), which has survived in its fullest re-telling in the rollicking (yet also esoteric) *Metamorphoses* of Apuleius, commonly referred to as the tale of "The Golden Ass" (which more properly should probably be termed "The Golden Tale of the Ass"). Earlier statues and mosaics show that this story of Eros and Psyche predates Apuleius by at least some centuries.

The story of Eros and Psyche occupies a very prominent position in the tale of Apuleius, who himself appears to have been an initiate of the Mysteries of Isis, and whose *Metamorphoses* should be read very carefully, because it appears to be the work of someone who understood the deep truths which the ancient myths were designed to convey.

The entire Tale of the Ass consists of stories within stories within stories, all of which related to us by a narrator who spends most of the book transformed into the outward appearance of an ass -- itself probably a metaphor relating to our incarnate experience in this material plane of existence.

In the story of Eros and Psyche (which I encourage all readers to read in the full original version provided by Apuleius), the beautiful maiden Psyche gets into trouble because she is so much more beautiful than all the other women on earth that she is compared to the goddess of beauty herself -- which not only incites the jealousy of her two older sisters but also upsets the proper order we discussed in a previous chapter, and earns Psyche the enmity of Venus (or Aphrodite) herself. Everything in this material realm has its source in the Invisible Realm, the realm of the gods -- and thus it would be improper to elevate physical beauty on this world to the level of the *source* of that beauty, the love goddess herself.

When no suitors dared to come to woo the beautiful maiden, her parents consulted the Oracle of the god Apollo, and received an answer that horrified them -- she was to be taken to a rocky crag while dressed in funereal garments, where she would be wed to a terrifying flaming serpent.

But, after the procession of mourning family and townspeople sorrowfully left Psyche on the cliff and returned home, a gentle wind carried her to a beautiful valley, containing a palace inhabited by invisible voices. She discovers that an unseen but loving husband comes and shares her bed each night, but strictly admonishes her never to try to see him and never to light a lamp in their room – and he always leaves before the morning sun.

However, to cut a long story short, Psyche becomes wracked by doubts (doubts which are originally sown by her two jealous sisters, who come to visit Psyche and whose jealousy grows even more ferocious when they see the sumptuous palace with its invisible attendants, and hear Psyche's story of her kind and loving but unseen husband). She decides she must see her husband's form, to find out if he is indeed a hideous monster, as her sisters confidently assert. And, impelled by their treacherous suggestions, she prepares a sharp blade in order to slay him, according to the plot laid out by the sisters.

Then come two "awakenings" in the tale. The first is the accidental awakening of Psyche's lover, as she bends over the bed to commit the foul deed of murder, and the lamp in her hand illumines not a horrifying monster but instead the handsome god of love, Eros (or Cupid), sleeping peacefully with his golden locks falling in curls over his chest, and the down upon his folded wings still moving in a "continuous running flutter."[99]

Overwhelmed by the sight, and the realization that her sisters were wrong, Psyche drops the blade -- but a drop of oil from the lamp falls upon the god's right shoulder, and in an instant he awakens and realizes that Psyche has betrayed her promises to him. Then he flies away.

Distraught, Psyche searches all over the earth for her lover, but cannot find him -- and (in the version as told by Apuleius) ends up in the palace of Venus herself, who was already completely opposed to Cupid's choice of a mortal woman as his bride, and who assigns a series of seemingly-impossible tasks to Psyche (this pattern, of course, is very familiar from the folklore and myths of the world). The final task involves an underworld journey -- to bring back a box containing Beauty from Proserpine (or Persephone), queen of the underworld.

Like Pandora, Psyche cannot resist the temptation to open the box and take a little for herself (again, a function of doubt -- in this case, of her own beauty, even though we have already been amply reminded that Psyche is the most beautiful of mortal women). But instead of containing Beauty, the box actually contains the Sleep of death -- and Psyche succumbs, until she is brought back to consciousness by the power of her divine lover Eros.

There are many layers of esoteric meaning in this story, of course -- but most pertinent to the theme of our investigation in this chapter is the lesson conveyed by the observation that Psyche (whose name itself could be seen as roughly synonymous with the "ego-mind") gives in to her doubts, and listens to the encouragement of her jealous and wicked older sisters, instead of listening to the counsel of the divine Eros himself, who loves her and desires what is best for her.

Like Thomas before his restoration in the encounter with the risen Lord, she is a picture of our own ego-mind, alienated from the divine Higher Self by virtue of her own doubts and wavering self-counsel (and her tendency to listen to the dubious suggestions of her wicked sisters, just as earlier in her life she listened uncritically to the adorations of the townspeople who said she was as beautiful as the goddess of beauty herself). But eventually, like Castor in the story of the Gemini Twins, Psyche is restored by the power of the divine partner in the story -- Eros, who awakens or revives her (a symbolic act), and then takes her with him to share again in his divinity, just as Polydeuces shares his own immortality with Castor.

Thus, we can agree that the Psyche story provides another powerful illustration of the centrality of the concept of the higher self in the world's myths.

Another even more practical example, perhaps, can be found in the character of Odysseus, in the Odyssey of ancient Greece. As I've argued at length in *Star Myths of the World, Volume Two*, Odysseus is uniquely a character in whom self-control as well as subtle attunement to the voice of the Higher Self are displayed to a remarkable degree, far beyond that of nearly any other character in the Greek myths.

The Odyssey contains countless examples in which Odysseus must rely upon his ability to hear the guidance from the Invisible Realm in order to survive a situation.

For instance, when he is first leaving Calypso's island after eight years of imprisonment with the amorous goddess, on a raft he has fashioned himself, and begins to make his way across the vast ocean, Odysseus is spied by the god Poseidon, who is furious to see the hero escaping and heading for home. Poseidon whips up a tremendous storm, with towering waves and howling winds, and quickly shatters the wayfarer's raft to bits.

Odysseus is initially rescued by the kindly goddess Leucotheia (or Leucothea), who herself had once been a mortal woman named Ino (in another manifestation of the message of the inner divinity of each man and woman). She gives him her sash for protection and advises him to strip off his clothes, tie her sash around his waist, and swim for the distant shore (the giving of the sash is another myth-pattern found in other Star Myths -- for instance in the story of Gawain and the Green Knight).

Odysseus spends three days swimming through the churning sea, and as he nears the forbidding shore, he perceives that the coastline is composed of rocky crags and cliffs, against which the massive billows are pounding, and finding nowhere to safely land, he fears for his life and begins debating back and forth in his mind, considering all the possible options.

He dares not try to land upon the sharp rocks, but then fears that if he tries to swim for a different spot he may be hauled back out to sea, or even devoured by a whale! The poem itself gives us a window into the thoughts of Odysseus as this entire debate is raging in his doubting mind:

> "Alas for me! since Jupiter [Zeus] has granted me to behold the unexpected land, and I indeed have passed over this wave, having cut through it, there has appeared no where any egress out of the hoary sea; for without there are sharp rocks, and a dashing surge rages around, and a smooth cliff runs up, and near it the sea is deep; and it is not possible to stand on both my feet, and to escape an evil plight, lest by chance a mighty billow snatching me, as I am going out, dash me against a stony cliff, and my attempt be in vain. But if I shall swim still farther, in case I may some where find sloping shores, and ports of the sea,

> I am afraid, lest the storm snatching me again should bear me to the fishy sea, mourning sadly, or even the deity should send a mighty whale against me, such as illustrious Amphitrite nourishes in great numbers: for I know how illustrious Neptune [Poseidon] is enraged against me."[100]

The ancient epic says that *even while Odysseus was turning over these considerations in his mind*, a mighty wave caught him and smashed him against the rocky coastline -- and he would have perished had not the goddess Athena "prompted his mind" and guided his actions.

It strikes me that this episode in Odyssey Book 5 powerfully dramatizes exactly the teaching we have seen the Katha Upanishad and the story of doubting Thomas and the story of Psyche trying to convey. It is not that Odysseus is not correct in carefully weighing the odds of landing successfully on the cliffs which are being pounded by the massive swells -- or that he is mistaken in determining that it would be suicidal of him to try to make his landing at this spot. But we see his mind start to wander further and further afield, until he is worrying about whales being sent to devour him, and starts saying to himself, "and I know that Amphitrite has numerous whales, and that Poseidon . . . " and as he is letting his mind run away with him, a great breaker snatches him up and hurls him against the rocky cliffs.

It is only by the intervention of the goddess Athena that he is able to survive this ordeal -- and the text itself says that the aid he receives from her is that she "prompted his mind" (in the literal translation of Theodore Alois Buckley published in 1896). A few lines later, the ancient poem tells us that Odysseus would have met his end there, "contrary to his fate," had not the flashing-eyed goddess Athena "given him prudence," and after that he swims parallel to the shore a ways, until he perceives a stream emptying out into the sea, and prays to the deity of that river to allow him to safely come ashore.[101]

This is very much akin to the encounter of Thomas and Jesus -- the doubting mind is stilled by the divinity, and the proper relationship is established. In the case of Thomas, he declares that he will be obedient to, or controlled by, his Lord. In the case of Odysseus, the goddess calms his mind, gives him prudence, and shortly thereafter we see the hero swimming confidently in the correct direction, with no further doubtful inner dialogue being recorded -- and he is soon safely on dry land again.

Note that the internal dialogue -- the mind's spinning and spinning upon itself – seems to be the activity that the goddess corrects in this instance. Odysseus is above all a character who is skilled with words -- and he uses those words to reason his way out of terrifying dilemmas throughout the Odyssey, and most of all to spin an unbelievable narrative for the people of Phaeacia, upon whose shores he has landed in the episode described above.

But even though his mind and his words can be a tremendous asset to Odysseus, it is quite evident in this particular example that his powerful internal narrative threatens to be his undoing. Throughout the epic, we find examples in which his control and balance are exemplary – and we find a few moments in which Odysseus apparently lets his ego-mind run away with him, and his passion get the better of him, leading to disaster (the most disastrous of these being his taunting of the Cyclops Polyphemus, as Odysseus and his companions are sailing safely away -- but Odysseus cannot resist declaring his name to the

Cyclops, resulting in the prayer to Poseidon that brings down the enmity of the sea-god).

This fact, if I am reading the message of the myth correctly, constitutes a tremendously powerful and practical lesson for our benefit here in this incarnate life. Our connection with the Infinite Realm, and with the guidance of the divine (such as the guidance Athena gives to Odysseus) is obstructed by the runaway words and internal "spinnings" of the ego-mind.

These endless mental narratives and debates are the source of the depiction of the ego-mind as being wracked with doubts, as is Thomas in the gospel narrative, or Psyche in the myth of Eros and Psyche. They can bring us to ruin -- as the doubt-narrative of Odysseus, dramatized in the Odyssey Book Five, illustrates in such memorable fashion. And they separate us from the Infinite Realm, because the Infinite Realm is pure possibility, and thus reposes in perfect stillness that is the exact opposite of our endless speculations over what outcome will manifest itself (which is exactly what the doubting mind of Odysseus begins to do, before Athena calms his mind).

As the opening line of the Tao Te Ching (or Dao De Jing) tells us:

道可道　　非常道

名可名　　非常名

"Way that can be uttered cannot be the constant (eternal) way; Name that can be named cannot be the eternal name."

It is only by stilling the endless utterings of our mind, and by quieting the endless "naming" of our internal monologue (or internal dialogue!) that we can be guided by the divine guide.

I would suggest that this stilling is one of the purposes of the practice of meditation, as well as the practice of ancient disciplines such as Yoga or of certain martial arts and internal arts which aim for the attainment of a state beyond the ego-mind's chattering.

And as we learn to listen to the voice of the ancient myths, we will find that they are teaching us the same thing.

The Universe Within

As we have seen, one of the keys to unlocking the treasures contained in the system of celestial metaphor underlying the ancient myths lies in understanding *not only* that the myths encode the cycles of the heavens, but also that the heavenly realms *themselves* point towards and allegorize the invisible and infinite realm of the gods.

The myths thus guide us towards the understanding that this realm exists, that it is vitally important, that it entwines and interpenetrates every aspect of the cosmos in which we find ourselves, and that it in fact supplies the very source and fountain of everything in the visible realm.

If we accept that this invisible realm exists and that it interpenetrates every aspect of the visible realm of our ordinary waking reality, then a moment's thought can lead us to the conclusion that the invisible realm must therefore connect us in some way to everything else in the universe -- that we ourselves are connected to the infinite realm, which is connected to everything else as well.

And this truth can be found among the teachings of the ancient myths.

For instance, the great cycle of the zodiac wheel, which forms the basis for the mighty "battles" in many ancient epics, can also be shown to have been overlaid in ancient times upon the human body, such that these great struggles can be understood to correspond not only to the endless interplay of light and dark that takes place in the great circle of the year, but also to the struggle between "higher" and "lower" in the human body itself, as we strive to elevate not only our consciousness but our vital energy upwards from the base energy centers to the crown of the head and beyond.

There is abundant evidence to support the assertion that the great battles recounted in the ancient epics of the Mahabharata (of ancient India) and the Iliad (of ancient Greece) embody the endless interplay between light and dark in the cycle of the year. As we've already seen, the year can be divided into two halves, during one of which light can be said to figuratively "dominate the battlefield" (when hours of light are longer than hours of darkness during each 24-hour period), and during the other of which darkness clearly "gets the upper hand" and the hours of darkness dominate during each 24-hour period.

Numerous clues in the Mahabharata point to the conclusion that the struggle between the Pandavas and the Kauravas upon the battlefield allegorizes the struggle between darkness and daylight throughout the year. In that battle, the Pandavas clearly represent the "upper half" of the year, during which daylight triumphs, and the Kauravas represent the "lower half" of the year, when hours of darkness outweigh the hours of daylight.

For example, the father of the Pandavas (Pandu) is a figure who has many characteristics which argue that he is a solar figure in the epic. For one thing, as argued in *Star Myths of the World, Volume One*, Pandu's two wives Kunti and Madri may have correspondences to the two planets which remain closest to the sun at all times, Venus and Mercury.[102] On the other hand, the father of the Kauravas, Dhritarastra, is blind -- which makes sense if his family represents the half of the year in which night dominates over day.

The correspondences are even more pronounced in the epic of the Iliad. Here, the Achaean forces who go *down* to Troy correspond to the upper half of the year. Their most illustrious warrior, Achilles, for example, is described as shining so brightly that one cannot bear to look at him, when he puts on the armor forged by the gods towards the climax of the action.

On the other hand, Hector is the greatest warrior of the Trojans. He is described as a "master of horses" (note that Sagittarius, the bow-wielding centaur or skilled horseman and bowman, is near the bottom of the zodiac wheel). Further, Hector wears a great black *horsehair* plume in his helmet. Much more analysis on the celestial foundations of action in the Iliad is available in *Star Myths of the World, Volume Two*.

One one level, then, these great battles, which have their foundations in celestial metaphor, can themselves be understood as emblematic of the interplay between light and dark throughout the year, but also between visible and invisible, higher and lower, spiritual and material.

But on another level they also point us towards an understanding of the same battlefield within our own human body, which reflects in a microcosm the infinite cosmos without. The battle between spirit and matter is metaphorically understood to play out as well between the "lower drives" centered at the base of our spine, near the "root chakra" and the genitals, and the "higher awareness" at the crown of our skull and opening into the celestial realms above.

The lowest chakras are rooted in earth, and the highest open up into the heavens. Both heaven and earth, of course, are equally indispensible and equally and infinitely important -- and yet the goal of the internal systems found in traditions such as kundalini meditation, Daoist *nei gong* and *nei daan*, Tantra, maithuna, and others involves the raising of the vital spiritual force upwards along this path from the lowest energy centers towards the highest.

For this reason, it is fascinating to note that the forces representative of the lower half of the year in the Iliad of ancient Greece -- the warriors of the Troy, led by Hector -- are associated with the city of Ilium, situated in the midst of the windswept plains. This ancient name of the city of Troy -- Ilium – is revealing, because the human body itself has a region known as the *ilium*: the wide-spread bone of the pelvis, the highest points of which are designated the ilium.[103]

And this is just one example of the way in which the myths and the mysteries they teach appear to be written upon the human body, as well as upon the vault of the stars (for the human body, as we will see, can be understood to reflect the infinite realm of the heavens).

The battle between the Achaeans and the Trojans, described in the Iliad, then, can also be understood as a battle to raise the vital energy from the lowest centers (dominated by our more basic, bodily drives) towards the higher centers (symbolic of the raising of the spirit nature, the divine spark, and the integration with our higher self).

Clearly, this struggle involves both the material and the spiritual natures in each man or woman. The metaphor-system of the myths portrays as one of its most central messages that the material and the spiritual are intertwined and entangled -- thus, we cannot separate the two, at least while we are here in this incarnate life. Our spiritual elevation will involve practices and disciplines which also have a physical component.

For example, the practice of meditation, which has as one of its goals the recovery of stillness which transcends the endless spinnings and rovings and doubtings and narrations of the ego-mind, may incorporate breathing techniques, body postures, and even hand positions (mudras). These physical details having to do with actions which involve the body are included in the teachings of the different meditation traditions, because the intertwined nature of the physical body and the invisible nature was recognized and understood by those who gave us these traditions. Even the position of the tongue within the mouth while meditating is considered significant in some schools of meditation.

The breath itself seems to have been understood to be a bridge or mediator between the physical and the spiritual. The very word *spirit* comes from the word meaning "to breathe" (the root *spir* having to do with physical *respiration* but also with *inspiration* that comes from another realm or plane of existence), and the same dual nature is found in other languages (the root *pneuma* seems to exhibit a similar duality, as do the concepts of *prana* and *vayu* in the ancient Sanskrit texts and language). And the word

which is pronounced *chi* in Mandarin Chinese (more frequently rendered into Latin letters as *qi*) and *hei* in Cantonese, also has a similar connection to both breath and to the invisible and internal vital energy.

When we understand that breath in some way connects body and spirit, then it is a short step to the understanding that singing and chanting must also involve the same intermediating role between the physical and the spiritual.

Chanting and singing adds the production of wave energy in the form of sound, as well as setting up resonance and vibration within our own physical frame and in the surrounding space.

The sacred traditions of cultures around the world have taught the importance of chanting and singing. Often, chanting is an important aspect of meditation practices. Notably, the recitation of mantras in the cultures of China and India and neighboring lands often involves the repetition of the mantra one hundred eight times (a precessional number), or other multiple which is connected to the celestial cycles.[104]

The chanting of the sacred syllable

usually rendered as *Om* or *Aum* in the Latin alphabet, presents an interesting study. The Yoga Sutras of Patañjāli recommend the chanting of the divine sound of Om as essential to the unveiling or revealing of the immutable self -- the supreme self -- as well as to the removing of all obstacles.

In the first of the four sutras of Patañjāli, we read:

> OM is a symbol for Ishvara [variously described or understood to be the Supreme Infinite, Supreme Being, and Source of All Knowledge].
>
> Repetition of OM [with this meaning] leads to contemplation.
>
> Through this practice, the immutable self is revealed and all obstacles [antaraya] are removed.[105]

Note that in the ancient teaching cited above, the repetition of the sound evoking the Supreme Infinite is described as leading to the revealing of the "immutable self" -- that is to say, the unchanging self, which cannot be a description of the ever-changing ego-mind represented by Thomas in John chapter 20.

It is when Thomas comes into proper relationship with the Christ that the tension and conflict is resolved: similarly, the Yoga Sutras teach us that the repetition of the sound of Om, resonating within our own body, can reveal to us our own immutable self.

The sound of Om (or Aum) has been explained to me as containing all the possible vowel sounds that the musical instrument of our body can produce, beginning at the lowest and deepest origin of vibration within our chest and lungs, and then moving up through entire column of sound -- through the throat, and then the chambers of the mouth and nasal cavities, until it reaches the very front of our mouth and is completed with the bilabial consonant of the "m" sound.[106]

In light of this explanation, it seems that chanting this sound is a way of evoking the Infinite using our body's capacity for creating vibration and sound -- as well as a way of elevating our material nature to greater integration with the infinite potential of the invisible or implicate realm. Remember that the metaphor of the invisible realm being the "seed realm" calls our attention to the fact that the realm of potential is unbounded -- it is not manifest and thus it has infinite potentiality. The repetition of a syllable which encompasses the entire range of vibratory sound which our *breath* (which mediates between matter and spirit, and which is connected with the concepts of prana and qi) can be seen as a declaration of this infinite and unbounded aspect of the infinite and divine nature, and of the infinite and divine realm.

When we utter this sacred syllable, in other words, (a syllable which "encompasses all," by its very nature) we ourselves are connecting with infinity, with unboundedness, with that realm of the gods in which all is possible.

Notably, when the meaning of the sacred sound Om is explored by B. K. S. Iyengar in *Light on Yoga*, he says:

> According to Śri Vinobā Bhāve, the Latin word Omne and the Sanskrit word Aum are both derived from the same root meaning all and both words convey the concepts of omniscience, omnipresence and omnipotence. Another word for Aum is praṇava, which is derived from the root nu meaning to praise, to which is added the prefix pra denoting superiority. The word, therefore, means the best praise or the best prayer.[107]

Thus, the world's Star Myths point us towards the reality of our connection with the Infinite, and towards the importance of our recognition of and integration with the Infinite, and in the ancient received traditions involving chanting, such as in the repetition of the sacred sound of Om but also in other traditions found worldwide, we perceive a connection between the man or woman and the Supreme Infinite.

It is very noteworthy to observe that the divine name in the Hebrew Scriptures similarly encompasses all the possible vowel sounds, although they are arranged differently. In a discussion of Plutarch's important treatise entitled "On the 'E' at Delphi," Robert K. G. Temple wrote in an appendix to his famous 1976 book *The Sirius Mystery* that the ancient understanding of the divine name involved one vowel sound for each of the eight notes of the musical octave, and that ancient oracle centers (such as Delphi) were placed on the earth at specific latitudes (eight successive latitudes), each assigned to one of these eight vowels (which itself was associated with a specific god).

In Appendix 4 of *The Sirius Mystery*, Temple writes:

> Demetrius of Phalerum, the student of Aristotle's Lyceum and who founded the famous great library of Alexandria when later in life he was exiled to Egypt, tells us in his surviving treatise On Style: 'In Egypt the priests sing hymns to the gods by uttering the seven vowels in succession, the sound of which produces as strong a musical impression on their hearers as if a flute and lyre were used.'

In Chapter XVI of *The White Goddess*, Robert Graves discusses this too, and there quotes Demetrius. Graves also refers to an eight-letter version of the sacred name. It may be that if one wants to count the base oracle center (which in musical analogy is the octave expression of the top centre) one should have an eight-letter version.

The version of the name is:

JEHUOVAO.

Note that E is the second letter.

We are faced with archaeological evidence that the second vowel, E, was prominently associated with the second oracle centre in descending order. And we know from Herodotus that Dodona, the top oracle centre, was said to be founded by Egyptian priestesses from Thebes in Egypt. We also known that certain Egyptian priests sang the seven vowels (or eight vowels, including an aspirate) in succession.

We have already seen that the geodetic oracle centres seem to have an octave structure. And as this book went to press a discovery became known which demonstrated the existence of the heptatonic, diatonic musical scale in the ancient Near East. We may even make a presumption that the uttering of the seven vowels in succession may possibly have corresponded to the seven notes of the octave (but we may never know that for certain).

And it is most important to emphasize that, however bizarre to us, the association of a vowel with an oracle centre is not our invention or surmise. The E may not only be read about in Plutarch but seen on ancient coins and on the omphalos stone itself. And this association of the second vowel with Delphi has never been explained by anyone.

So granted all the above, what follows? If each oracle centre had a vowel associated with it, then the second vowel being associated with the second centre would seem to imply a corresponding arrangement for the other centres. And if that is the case, it would seem that the entire system would be associated with and actually comprise a geodetic spelling-out, over eight degrees of latitude, of the unspeakable holy name of God, known commonly to the Hebrews as 'Jehovah.'[108]

This argument is strengthened by the undeniable fact that the name of the god *Jove* or *Iove* is clearly linguistically related to this same sacred holy name composed of all of the vowels.

And, astonishingly enough, it also cannot be denied that this same holy name was known among the Native American peoples prior to the arrival of Columbus. As has been previously mentioned in earlier books, the accounts of the numerous Hopi elders interviewed by Frank Waters and Oswald White Bear Fredericks give the name of one powerful creator god as *Taiowa* or *Ta-Iowa*, which is nearly identical to the sacred name in the Hebrew Scriptures.[109] Further, we have the evidence staring out at us from any map of the modern United States that one state has been named after a version of this holy name which was preserved in the tradition of the indigenous nations: *Iowa.*

An additional point forwarded by Robert Temple in a different part of *The Sirius Mystery* also bears mentioning during this discussion, and that is his observation of an assertion by Godfrey Higgins (1772 - 1833, whose *Anacalypsis* was published in 1836 and constitutes another important and influential exploration of the commonality of the world's mythology) that the word *omphalos,* designating the "navel stone" found at Delphi and later applied to similar stones found at other important sacred sites around the globe, has connections to "the sacred syllable *om* of the Indo-Europeans."[110]

In these arguments from Robert Temple (and Robert Graves, and Godfrey Higgins), we are beginning to see a connection not only to the human body (in the chanting or uttering of the sacred sound) but also to the planet earth itself -- upon which oracle centers were established in successive degrees of latitude (as Temple demonstrates), some of which contained omphalos stones which may also connect back to the same all-encompassing syllable discussed above.

Connections between the *microcosm* of the individual man or woman and the *macrocosm* of the external universe, including our own planet earth as well as the heavenly actors of sun, moon, stars and planets, abound in ancient tradition and myth.

We have already observed the connection between the human body and the actors and events in the Iliad (which themselves mirror celestial figures and celestial cycles), explicit in the naming of the ilium within our own skeletal system. Other ready examples include the crossing of the Red Sea within the book of Exodus, which I have shown to involve the crossing of the Milky Way band between Scorpio and Sagittarius -- corresponding to the location of the great turning point of the year -- and which Alvin Boyd Kuhn has argued can also be understood as a metaphor for this incarnate life, which we indeed traverse within a "red sea" of sorts: the sea of our own blood, which gives us life in this body.[111]

Another example is the encounter with the Infinite described in the episode of "Jacob's Ladder" (the vision at Bethel). I have also shown that this episode is almost certainly patterned after the region of the shining Milky Way band between Sagittarius and Scorpio, along with Ophiucus and Hercules.[112]

Depiction of Jacob's Vision, from 1860. Jacob's outline corresponds to Scorpio, the depiction of the Almighty corresponds to Hercules, and the stairway or "ladder" corresponds to the column of the Milky Way galaxy. The mount between the trees above Jacob's face corresponds to Ophiucus.

Once again, however, there are also connections to our mortal physical body, through which our soul makes its sojourn in this material realm on its own stairway of elevation. It has been pointed out by the insightful advocate of astrotheology and the unity of the world's sacred traditions, Santos Bonacci ("Mr. Astrotheology"), that the ladder between the lower realm of earth and the celestial realm of heaven can also be understood to signify our own spinal column, which reaches between the lower plane of our root chakra region (and the plains of Ilium) to the cervical bones and the skull itself -- which resembles the very vault of heaven, and which contains our brain matter, through which the infinite universe is infused as we manifest consciousness in our current form.[113]

In a lecture on this subject, Santos points out that in the Genesis 28 passage, Jacob rests his head upon a rock or upon stones (Genesis 28: 11), and that the lowest bones of the spinal column are fused together into the sacral bone or sacrum -- which Santos quite convincingly argues could correspond to the stone or stones in the text.[114]

The connection to the human body in these ancient sacred texts and myths is undeniable.

And the connection is not just some kind of frivolous literary or poetic exercise -- it is intended to convey many layers of deep knowledge, for our benefit in this incarnate life. The connection has significance for our physical and spiritual health.

The vision of the individual man or woman as a *microcosm* -- a "miniature cosmos," reflecting and containing the same infinite universe we see around us -- is described most explicitly in some of the traditions preserved in ancient India and ancient

China (where the Empire of Rome never conquered during ancient times).

One ancient text, probably in existence by the second century BC but (in my opinion, although there are scholars in the field who will disagree with this assertion) likely preserving a body of wisdom which was already well-developed by the time of its composition and thus much older, is the text known as the

黃帝內經

which is pronounced *Huang Di Nei Jing* (or *Wong Dei Noi Ging* in Cantonese) and which signifies "Yellow Deified-Ancestor Internal Classic" -- the *Yellow Thearch's Classic Text of Internal Medicine*.[115]

It consists of two main sections, but in each we find Huang Di conversing with one of his ministers or advisors, named Qi Bo. The subject concerns health, longevity, and especially the qi -- its cultivation, its preservation, its circulation, and the techniques for correcting health issues, which at their root are understood to concern the flow of vital energy through the body.

The specifics explored in the text are beyond the scope of this chapter: what concerns us in this discussion is the fact that the worldview of the Nei Jing clearly assumes a deep connection between human body's energy fields and the cycles of the earth and heavens -- in other words, between microcosm and macrocosm.

For example, the text explicitly describes the energy of the four seasons as "the basis of death and life" and says that opposing the seasons leads to catastrophe, while following them enables one to achieve the Way (the Dao).¹¹⁶ The connection between the individual and the seasons is so important, in fact, that one is advised to regulate the patterns of sleeping and waking in accordance with them: rising early in the morning during spring, summer, and autumn but rising late during the winter; retiring to bed late in the spring and summer but early in both autumn and winter.¹¹⁷

Obviously, if the seasons are caused by the earth's position relative to the sun (and specifically the orientation of the axis relative to the sun), then our regulation of waking and sleeping in accordance with the seasons constitutes the ordering of our own motions in accordance with the motions of our planet earth.

That advice may not seem so surprising or significant -- but it is just the beginning of the ways in which the Nei Jing reveals an understanding of the harmony between the "inner cosmos" of the individual man or woman and the universe outside. The ancient text describes a connection between specific internal organs and the cardinal directions which relate to the orientation of our planet itself -- the liver being associated with east, for example, the heart with the south, and the kidneys with the north.¹¹⁸

The Nei Jing also describes a connection between major internal organs and the five visible planets -- with the liver having a connection to Jupiter, the heart to Mars, the "center" of the body (perhaps the spleen) to Saturn, the lungs to Venus, and the kidneys to Mercury.¹¹⁹ Clearly, this level of perceived correspondence indicates a vision that we can describe in terms of the microcosm and macrocosm. The understanding that informs the Nei Jing undeniably envisions the human body as connected to, and reflecting internally, the infinite realm around and above us.

The Nei Jing even lays out the five flavors which govern the food, the various combinations of which were understood to have either beneficial or detrimental impacts on the internal qi and the overall health.[120] Intriguingly, each of these five flavors is explicitly linked to one of the organs and thus (potentially) to one of the five visible planets as well. The flavors are sour (matched with the liver), bitter (matched with the heart), sweet (matched with the spleen), pungent or acrid (matched with the lungs), and salty (matched with the kidneys).[121]

Elsewhere in the Nei Jing (specifically in the section called the *Ling Shu*, which deals with needle therapy or "acupuncture"), we are told that the human body has "twelve major channels and three hundred sixty-five pathways" through which vital energy flows and upon which the practitioner of healing with the needles must focus. Just as we saw with the connection of the organs to the visible planets, these numbers 12 and 365 almost certainly relate to the cycle of the earth's annual orbit around the sun (passing through the twelve stations of the zodiac or the twelve months, and rotating on its axis approximately 365 times as it makes its circuit).[122] This understanding of the arrangement of the channels or meridians within the human body thus constitutes another evidence of a view of our physical form as a microcosm of the universe around us.

Nor do we have to speculate on the question of whether or not the numbering of the channels and pathways indicates a vision of the body as a reflection of our planet which has its own meridians and energy lines -- the naming of the various channels and acupuncture points themselves frequently reveals the same explicit understanding, with the Ling Shu calling some of them "streams," some of them "seas," and some of them "mounds" and "mountains."

There is even a most noteworthy metaphor used in the Nei Jing (and in particular the Ling Shu) which compares the relationship between an individual's good qi energy and an invasive and potentially disease-causing evil qi to that of a host and a guest or visitor, with the individual's good qi envisioned as the host and the evil qi described as the guest or visitor.[123] The text of the Ling Shu seems to imply that if the host is too weak, or does not take care, the guest can become a "bad guest" that takes over the mastery of the household from its proper owner.

The reason this particular comparison is so notable is that other ancient traditions, including those involving astrotheological characters and events, appear to make use of this very same metaphor.

For example, the theme of good and bad guests runs through the Odyssey and forms one of the central themes of that epic. The suitors who have invaded the home of Penelope, Telemachus and the absent Odysseus clearly play the role of "bad guests" -- but variations on this same relationship are explored throughout the text, from the imprisonment of Odysseus on the island of Calypso, to his account of encounters with the Lotus-Eaters and the island of the cattle of the Sun and the formidable goddess Circe and especially the Cyclops, Polyphemus (to name just a few). Sometimes we see good hosts and bad guests, sometimes bad hosts and good guests, sometimes good hosts and good guests -- a wide range of possibilities are presented for consideration.

The use of this same relationship to depict the condition of the internal energy of the body in relation to the threats to health raises the possibility that Star Myths such as the Odyssey might be teaching knowledge about the human body in addition to all the spiritual messages that we can unlock when we begin to understand its celestial correspondences -- and it is my contention that, in the view of the framers of the ancient wisdom, matters of spirit and matters having to do with the body were seen as closely and intimately related (while we are in this incarnate life, at least).

Note also that the imagery of the householder being overwhelmed by the evil visitor(s) is also explicitly present in the gospel accounts and the teachings ascribed to Jesus, in which at one point he declares:

> When the unclean spirit is gone out of a man, he walketh through dry places, seeking rest, and findeth none.
> Then he saith, I will return into my house from whence I came out; and when he is come, he findeth *it* empty, swept, and garnished.
> Then goeth he, and taketh with himself seven other spirits more wicked than himself, and they enter in and dwell there: and the last *state* of that man is worse than the first. Even so shall it be also unto this wicked generation.[124]

Clearly, there seem to be parallels in this teaching to the teachings on preventing (or, if necessary, removing) evil qi in the Ling Shu, which is also allegorized as a bad guest taking over. Is it possible that some of the hidden layers of meaning in the spiritual traditions from some cultures are made explicit in traditions and texts that have been preserved in other cultures?

I believe it is very possible, even likely.

I also believe it is possible that the understanding of other surviving Daoist disciplines involving the inculcation, fortification, and elevation of the vital energy and ultimately the spirit could help throw light on some of the teachings preserved in the Biblical scriptures involving sexual practices and sexual prohibitions -- which have been cast in a moralistic light by centuries of literalist influence but which may have more to do with practices akin to internal alchemy.

For instance, there is the well-known story in Genesis 38 of the fatal outcome of Onan's decision to spill his seed or semen onto the ground, which displeased the LORD and resulted in Onan's death (Genesis 38: 9 - 10). It is at least a possibility that this passage encodes teachings about sexual intercourse without ejaculation which is the subject of ancient Daoist teachings on sexual intercourse which are preserved in the fragments of certain ancient texts, the original titles of which are not known but which include texts known now as the "Classic of the Secret Methods of the Plain Girl," the "Recipes of the Plain Girl," the "Secret Prescriptions for the Bedchamber," and the "Principles of Nurturing Life" (among others).[125]

In his 1961 study *Sexual Life in Ancient China*, Robert van Gulik cites the research of the early twentieth-century scholar Yeh Tê-hui (1864 - 1927), whose reconstruction of surviving quotations and contents of the ancient books suggests that they all followed a similar pattern, always beginning with a chapter containing "Introductory remarks on the cosmic significance of the sexual union, and its importance for the health of both partners."[126]

Once again this pattern in the ancient texts reinforces the connection between macrocosm and microcosm -- the universe outside and the universe within.

Further, it can also be argued that the remainder of the chapter in Genesis which recounts the fate of Onan employs celestial metaphor, based on the presence of a kid, sheep shearing, and other details in the account which could indicate that the events which follow from verse 11 through the end of Genesis 38 may have a celestial foundation, most likely in the vicinity of the zodiac constellations Capricorn and Sagittarius, and the Milky Way band.

Thus, this rather sexually explicit chapter from Genesis would also fit into the pattern identified by Yeh Tê-hui and Robert van Gulik of ancient texts which connect human sexual activity with the motions of the cosmos.

What are we to conclude from all of the diverse evidence presented in this chapter from the ancient wisdom preserved in different forms in different cultures around our globe?

I believe that one likely conclusion would be that humanity has been given a plethora of disciplines and practices to enable us to pursue one of our central purposes in this incarnate life: the elevation of the vital spirit and the integration with the divine higher self. These disciplines include, but are not limited to, practices such as Yoga, meditation, chanting, breathing practices found in various cultures, inner work or internal alchemy, certain martial arts, and even sexual practices preserved in traditions such as the Daoist bedroom arts, maithuna, and Tantra.

These practices are all informed by the vision of the individual man or woman as a reflection of the wider cosmos -- containing the infinite universe within, underneath the dome of the skull which resembles and reflects the vault of the heavens (which itself is of course infinite).

I would further submit that these types of disciplines, practiced diligently under the care of an advanced teacher for many years, may perhaps correspond to the "celestial weapons" which are described in ancient texts such as Mahabharata -- weapons which Arjuna brings back after traveling to the realm of the gods, and which are essential to his success on the battlefield of Kurukshetra (itself a metaphor for the "arduous struggle" of this incarnate life).

And we ourselves can also, like Arjuna, travel to the celestial realms to obtain the assistance we need in our own Kurukshetra -- because we have access to the heavenly realms, through the fact which is undeniably demonstrated in the world's ancient myths and sacred teachings: that we are inseparably connected to the infinite realm, through the universe which is right inside of us all the time.

Humanity's Ancient History

What are the ramifications arising from the overwhelming evidence of the existence of an ancient world-wide system of esoteric metaphor which forms the foundation upon which virtually all the ancient myths and sacred traditions held by cultures separated by vast distances and even by great gulfs of time?

The implications of such a connection are enormous -- and they rock conventional theories of humanity's ancient history to the core.

For decades, brave researchers have written and spoken about the evidence around the globe which points to the possibility of an ancient, advanced civilization forgotten or ignored by the conventional paradigm of human history -- one which appears to have predated ancient Egypt, ancient Mesopotamia, ancient China, and even the ancient Indus-Saraswati culture.

Pioneering researchers such as John Anthony West, Robert Schoch, Graham Hancock, Christopher Dunn, Joseph Farrell, John Michell and many others -- stretching back to Hertha von Dechend and Giorgio de Santillana, as well as to researchers of previous decades in the twentieth century such as Alfred Watkins, Katharine Maltwood, and R. A. Schwaller de Lubicz -- have provided volumes upon volumes of evidence and analysis and arguments that shows that the conventional paradigm of human history is gravely flawed. And that is to say nothing of the many scholars, "antiquarians," and other thoughtful men and women of previous centuries who perceived the same thing and wrote or spoke about it previous to the twentieth and the twenty-first.

For years, these arguments were dismissed and ridiculed, despite massive evidence from existing monuments around the world, evidence such as the weathering on the Sphinx of Giza and its surrounding Sphinx Enclosure, as argued by John Anthony West and Robert Schoch based on extensive analysis beginning in the late 1980s, or the fact that monuments around the world can be shown to be located at significant intervals of longitude corresponding to precessional numbers, as demonstrated by Graham Hancock, or upon great circle lines stretching across vast distances and across oceans, as demonstrated by Jim Alison – to name just a few of the most difficult to dismiss, among literally thousands of other interconnecting pieces of evidence and lines of argument.[127]

Part of the rejection of the evidence rested upon the fact that the ancient monuments on our planet are dated by conventional arguments to cultures and civilizations operating within known periods, such as the above-named Sphinx (argued to be the product of dynastic Egypt, although the work of John West and Robert Schoch poses serious challenges for that assumption).

While ignoring or downplaying the evidence that creating the precision stonework found in many ancient sites and artifacts (such as the stones at Puma Punku, or at Baalbek) or siting these monuments along great circle lines or at precise longitudinal intervals would all require knowledge and abilities difficult to explain in the conventional framework, the defenders of the academic paradigm could remain on offense instead of defense by pointing to the lack of archaeological sites unquestionably older than the timeline of known civilizations, and asking proponents to produce hard archaeological evidence of this lost ancient civilization.

As Professor Robert Schoch vividly remembers, and as was reported in the *New York Times* when it happened, Egyptologist Mark Lehner publicly attacked Robert Schoch's proposition that the Sphinx could have been constructed prior to 2,500 BC, calling Professor Schoch's work "pseudoscience" and asking, "If the Sphinx was built by an earlier culture, where is the evidence of that civilization? Where are the pottery shards? People during that age were hunters and gatherers. They didn't build cities."[128]

The excavation of the site now known as Göbekli Tepe, located in modern-day Turkey near the border of Syria, changed the game completely. Excavation began in 1995 by a team of archaeologists led by the late Klaus Schmidt.

Göbekli Tepe is a massive complex, only a small portion of which has been excavated. Robert Schoch describes its scope by saying, "Picture Stonehenge, multiply it by twenty, carve the pillars more ornately, place the circles next to one another, and intentionally bury them with a mountain of rock and dirt."[129] Finely carved T-shaped pillars of limestone standing as tall as eighteen feet and weighing up to perhaps ten or fifteen tons are arranged in circles with diameters that range from around thirty feet to nearly one hundred feet.[130] While only a handful of these circles have been excavated so far, Professor Schoch notes that geophysical surveys indicate the entire complex may cover an area of up to ninety meters, and contain another sixteen to twenty circles which remain completely untouched to date![131]

These pillars and circles, and the beauty and sophistication of their workmanship and their artwork (in both low relief and high relief), as well as their size and scope are astonishing enough -- but even more astonishing discoveries began to come to light as the excavations proceeded.

For one thing, it is evident that the entire complex was carefully and deliberately buried under thousands of tons of rock and earth -- a massive task -- for reasons unknown.

For another, the burial of the site enables radiocarbon dating and other forms of dating on the site, including dating on the burial of the site, which must have taken place some time after the site was originally constructed -- and the dating that has been done indicates that the site was buried not later than about the year we call 8000 BC (or BCE).[132] The site itself may date back to 9000 BC, or perhaps even 10,000 BC, suggesting that it was built up to 2000 years before it was buried (think back to what was going on 2000 years ago from today's date, to get some perspective on the antiquity the site may have had before it was buried).

The incredible antiquity of these results completely upends the conventional model of human history. The data suggests that the site was *buried* by 8000 BC, which is about 5,500 years before the supposed 2,500 BC date of construction of the Sphinx at Giza held by conventional historians.

Note for perspective that 2,500 BC, when the Sphinx at Giza was supposedly designed, is "only" about 4,500 years ago from our modern point in history, in the early twenty-first century. In other words, the time of the *burial* of Göbekli Tepe was already further back in history at the time of the Old Kingdom of Ancient Egypt than the Old Kingdom of Ancient Egypt is removed from our own moment in history (further back by a full 1,000 years)!

And the time of the construction of Göbekli Tepe may have been another 1,000 years earlier than that, or even 2,000 years earlier.

The size and scope of this massive monumental site, as well as the workmanship and quality of the pillars themselves, not to mention some of the beautiful and sophisticated artifacts such as jewelry and statuary which has also been found in the site, indicate that the Göbekli Tepe culture had achieved masterful levels of engineering and manufacturing skill. It is a site which attests to a sophisticated culture or civilization which had the resources and security to develop extremely specialized and advanced stoneworking techniques and to execute them on a massive scale, all with a level of delicacy and artistic beauty which is astonishing. And it is a site which attests to the existence of such a culture at a date that completely upends the conventional timeline of ancient human history.

In short, the existence of Göbekli Tepe argues for the existence of exactly the kind of "lost civilization(s)" or forgotten predecessor culture(s) that researchers such as Graham Hancock, Robert Schoch, John Anthony West, Joseph P. Farrell, and many others have been proposing -- and proposing since the early 1990s, before Göbekli Tepe was even excavated or dated.

While of a very different sort of evidence from the stone pillars and datable burial material of Göbekli Tepe, I would propose that the extensive and at this point virtually undeniable evidence found in the world's myths, scriptures and sacred stories -- including the myths and references found in the earliest Pyramid Texts and Mesopotamian tablets and Vedic scriptures -- showing the existence of a common, unified, world-wide foundation of celestial metaphor (in which many of the very same distinctive and sometimes obscure references are found again and again, in cultures from around the globe) may form yet another very strong set of evidence pointing to the existence of an extremely ancient, very sophisticated, and now forgotten culture or civilization, and one which also must have predated the ancient civilizations of Mesopotamia, of Egypt, of China and of the Indus and Saraswati River regions.

Both Graham Hancock and Robert Schoch have also found extensive evidence that the forgotten ancient civilization or civilizations were *destroyed by cataclysmic events* in remote antiquity, around the period of roughly 9600 BC, although they do not necessarily agree on the mechanism that caused the disaster, with Graham Hancock arguing for the possibility of a comet strike by a massive comet which disintegrated into multiple smaller bodies before impacting the earth across a tremendous area, and Robert Schoch arguing for the possibility of a massive outburst of solar radiation which would have ended the ice age in a matter of days or weeks and may have even made the surface of the earth radioactive for some period thereafter.

Whatever the mechanism, both researchers point to evidence of cataclysmic forces at work on our planet around the same time period, as well as to evidence that the survivors of this catastrophe may well have retreated to extensive underground cities and tunnels in order to escape the devastating effects -- perhaps to re-emerge many generations later to try to rebuild and restore some semblance of what had been lost.

The very ancient civilizations of Egypt, Mesopotamia, and the Indus-Saraswati may be representatives of the first reemergence, long after the destruction — and may have preserved memories and technologies of the civilization(s) and culture(s) that had vanished so many thousands of years before.

It may be that the world's ancient Star Myths, and the worldwide system of astrotheology upon which they are based, also testifies to the memory of a system which was already incredibly ancient by the time the Pyramid Texts or the Vedas or the Gilgamesh tablets were written down.

Strikingly, the pillars of Göbekli Tepe appear to have astronomical alignments, as Robert Schoch has argued in *Forgotten Civilization* (2012).[133] In that book, presents the inspired analysis that some of the pillars themselves may represent celestial figures, based upon the fact that arms and hands are carved in low relief on some pillars, and some appear to have loincloths suggestive of the pelts of animals. Based on these details, and the presence of a dog-like or fox-like animal on a pillar with arms (but no head), Robert Schoch suggests the possibility of a connection to the constellation Orion -- the "headless hunter" in the sky (the head of the constellation in the sky is indicated by extremely faint stars).[134]

Even more astonishing, however, are the arguments presented in Graham Hancock's more recent *Magicians of the Gods*, published in 2015, making the case that the figures carved in relief upon Pillar 43 in Enclosure D at Göbekli Tepe correspond to the constellations Sagittarius, Scorpio and Ophiucus, and that the disc depicted off the wing of the condor-like bird corresponding to the "bow" or "teapot" formation in Sagittarius represents the solar disc, crossing the Great Rift in the Galactic Core of the Milky Way band, between Sagittarius and Scorpio.[135]

Graham bases these arguments on insights presented in a paper written by Paul Burley in June of 2011, as well as upon Graham's own visits to Göbekli Tepe and his discussions with Paul Burley, as he describes in the book *Magicians of the Gods*.[136]

Adding his own analysis to the observations of Paul Burley, Graham makes the brilliant suggestion that the shape of the vulture-figure could correspond to the "teapot" section of stars in Sagittarius -- which it in fact fits very satisfactorily. He notes that the bird seems to fit the stars even better than the traditional outline of Sagittarius fits them -- but note that he is apparently not using the outline-method suggested by H. A. Rey, but rather the same kind of elaborate and flowery Sagittarius artwork that H. A. Rey set out to correct.[137]

Based on the actual orientation of the stones in Enclosure D, and the visibility of the sky from its location, Graham advances the argument that the event depicted on this particular stone points to the time when the winter solstice sun is passing through the region of the Galactic Rift, between Sagittarius and Scorpio – as it is at present, and as the Maya Long Count calendar famously also commemorates, with its countdown which ended at winter solstice in 2012.[138] These astonishing connections are bolstered by the undeniable similarities which Graham Hancock demonstrates – along with full-color photographs by Santha Faiia -- between reliefs found on the stones of Göbekli Tepe and reliefs found at sites in Cutimbo and Cuzco in Peru, on the western side of South America (pointing to potential very ancient contact with civilizations or cultures in the Americas, just as Professor Schoch's book notes potential connections to the monumental moai of Rapa Nui in the Pacific).

These archaeological discoveries are absolutely fatal to the conventional narrative of humanity's ancient history. They point overwhelmingly to the existence of an advanced but now forgotten ancient culture, a culture capable of marvelous feats of engineering, and a culture possessed of extremely sophisticated knowledge of the heavenly cycles, cycles which it imbued with profound significance.

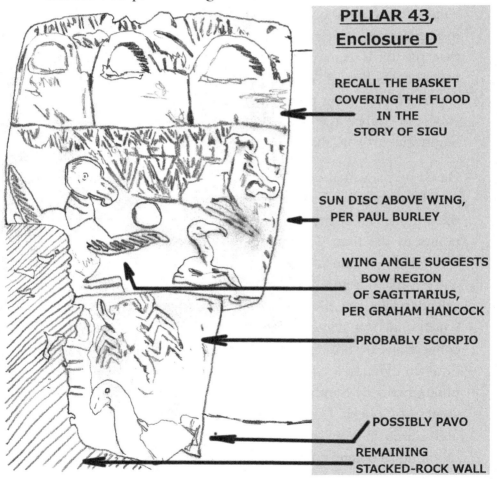

Pillar 43 from Enclosure D. Note the bend in the wings of the "vulture" next to the disc: these angles are strongly suggestive of the outline of the bow in Sagittarius as outlined by H. A. Rey.

To all of this archaeological evidence we can now add *mythological* evidence which demonstrates beyond doubt that the world's ancient myths, scriptures and sacred traditions are built upon a *common system* of celestial metaphor. Like the archaeological evidence, this mythological evidence is worldwide -- with distinct similarities that are undeniable even across vast distances and within cultures separated by great gulfs of time.

Returning again to the haunting words found in the introduction to *Hamlet's Mill*, both the physical evidence and the mythological evidence indicates that we are indeed standing in the midst of the ruined remains of a "great worldwide archaic construction," one so old that "the dust of centuries" already lay thick upon it by the time the ancient Greeks "came upon the scene."[38]

We have been taught by the conventional narrative that all these ruined components, scattered around the planet, are completely isolated from one another -- but in fact, as the authors of *Hamlet's Mill* suggest (and as researchers in the decades since have all but proven), the pieces in reality represent "fragments of a lost whole." As *Hamlet's Mill* says, "they make one think of those 'mist landscapes' of which Chinese painters are masters, which show here a rock, here a gable, there the tip of a tree, and leave the rest to imagination."

As in those paintings, vast portions of the archaic construction are no longer visible to our sight, but they are all in some way connected. And yet, for some reason, the keepers of the conventional timeline insist that any proposed connection is preposterous. In doing so, they invite those who accept their paradigm to see humanity as divided and disjointed. It is as if someone were to stand on one little "island in the mist" in one of those Chinese paintings that *Hamlet's Mill* describes, and declare that they are part of their own separate landscape, and have no connection to what is going on over on the other side of the mist-covered areas within the same painting!

And yet those living in ancient times demonstrate in numerous places within their surviving writings an understanding that all the myths are actually connected in some way. We can see this whenever a writer (such as Plutarch, or Plato) declares that a god or goddess in a different culture (such as Egypt) corresponds to a god or goddess in his or her own culture -- saying, for instance, that Thoth is Hermes, or that Amun is Zeus or that Set is Typhon.

The false paradigm of division and disconnectedness appears to be a product of literalist interpretation of the scriptures that make up what has come to be known as the Bible -- and this particular version of self-isolating Biblical literalism appears to have taken off in the centuries we call the 2nd through the 5th centuries AD. That is also when the Oracle at Delphi and the Eleusinian rites were shut down, under the Roman Emperor Theodosius (when the Empire had been declared Christian).

Of course, it is evident that much of the ancient wisdom was already in fragments by the time of the Greeks and the Romans -- perhaps as a result of the cataclysm that ended the now-forgotten civilization. The rise of the intolerant Biblical literalism, which not only declared itself separate from the other versions of the ancient celestial system found in other Star Myths in other cultures, but also declared those versions to be heretical and worthy of elimination, led to a further separation from the remnants of that ancient wisdom.

I am convinced that the world's ancient myths, scriptures and sacred stories provide powerful evidence that the ancient culture or cultures which may have predated the ancient Egyptian, Mesopotamian, Chinese, American, Pacific, and Indus-Saraswati Valley civilizations by thousands of years possessed not only sophisticated *technological* capabilities and advanced astronomical knowledge but also profound *spiritual* wisdom.

It is fairly evident, from the arguments presented by Robert Schoch, Graham Hancock, Randall Carlson, and others, that these ancient cultures or civilizations met a sudden and catastrophic end -- but also that survivors of that catastrophe were able to preserve remnants of their knowledge until it could grow and flower again, millennia later, and that all human cultures around the world appear to have inherited aspects of that ancient wisdom.

I believe that it is also fairly evident, based upon arguments presented in my 2014 book *The Undying Stars*, as well as arguments presented by researchers and authors such as Flavio Barbiero, Timothy Freke, Peter Gandy, Joseph Atwill and others, that the literalistic, exclusive and monotheistic interpretations of some of these ancient scriptures marked a decisive break with the interpretations that were preserved in other cultures -- and eventually resulted in an aggressive destruction of many of those other cultures and the loss of the streams of ancient wisdom that they may still have preserved.

This destruction began within the boundaries of the lands which had been conquered by the Roman Empire, and included the suppression of the texts and teachings of the Gnostics who in many cases used the very same scriptures and stories but with a different lens of interpretation, and spread throughout western Europe, parts of Africa and Asia, and later (largely through European imperialism and colonialism) spread their suppression of the ancient myths and traditions to other parts of the world as well, including India, the Americas, the islands of the Pacific, Australia, and additional parts of Africa and Asia.

The driving force of this suppression, or at least an extremely important aspect of it, was the engine of literalistic Christianity in its many forms -- almost all of which declare the supreme importance of a literal and historical figure named Jesus who is both God and man, belief in whom is deemed essential to salvation, and to whom his followers will tolerate no rivals.

This belief-system provided (and still provides) the supposed "intellectual cover" and supposed "moral justification" for the colonization of other cultures, the seizure of their land and wealth and resources, and the forced conversion of their people to literalistic Christian teachings -- often by the forced indoctrination of their children and younger generations in schools and the prohibition of their traditional spiritual practices.

The conversion process has also historically involved the destruction of traditional images and artwork, the suppression of shamanic and ecstatic techniques for accessing the Other Realm (including the destruction and outlawing of drums and other aids for inducing ecstatic states), the suppression and outlawing of plants and substances which may assist in opening pathways to non-ordinary reality (sometimes called "entheogens"), and even the wholesale burning of texts containing ancient wisdom, such as apparently took place in the Americas with the deliberate destruction of Maya codices.

All of this, of course, constitutes a tremendous crime against those individual men, women and children who were affected -- and indeed against all of humanity, because we have all been impacted in some way by these actions. In fact, the shape of the world today bears the enormous and hideous handprint of western imperialism on virtually every shore and on every continent and island. This is not to deny the good things which arose within western European culture in general, including artistic, philosophical and technological advances and developments, but rather to argue that those good things could have been accomplished without oppressing other cultures, stealing their resources, destroying their traditions and spiritual practices, enslaving their people, and establishing systems and structures to keep them from realizing their own artistic, philosophical and technological advances.

Additionally, the tremendous and bitter irony is that the exact same system of ancient wisdom which underlies the stories and characters found in the Bible can be shown to underlie and inform the myths and sacred traditions of all these other cultures and civilizations around the world -- and to teach principles which are very much the opposite of the intolerant and exclusive teachings found in many forms of literalistic Christianity.

By literalizing and externalizing the stories in the scriptures of the Bible, their ancient meaning has actually been inverted and turned into the opposite of what they were intended to convey. This inversion can be demonstrated in numerous instances -- perhaps none of them more clearly than in the story of Shem, Ham and Japheth, the three sons of Noah who after the flood find their father passed-out drunk and naked (or at least "uncovered") and whose reaction leads to either their blessing (in the case of Shem and Japheth, who cover him up by walking backwards with a sheet between them) or cursing (in the case of Ham, who ridicules and belittles their father in his embarrassed condition).

It is a story that vividly contrasts blessing and cursing -- lifting up the spiritual and divine in oneself and in one's fellow men and women, in the case of blessing, or beating them down and reducing them to the material, to the physical, to the body, and ultimately to a corpse, in the case of cursing.

And it is a story that has unfortunately been used by literalistic interpreters to argue that some peoples and ethnicities on our planet are literally descended from Shem, some from Japheth, and some from Ham -- and that those descended literally from Ham share in the curse affixed to him in the ancient story. This literalistic interpretation has wrongly and tragically been used to "justify" the oppression and even enslavement of men and women who are supposedly descended from the "race of Ham." You can still find without too much trouble maps from previous centuries (including as recent as the nineteenth and even the twentieth) labeling peoples on various parts of the globe as being descendants of these three sons of Noah.

However, as I have shown in many places before (including in an extended discussion with numerous diagrams in *Star Myths of the World, Volume Three*, as well as in my blog and website online), the story of Shem, Ham and Japheth and the drunkenness of their father Noah is built upon celestial allegory, and it involves the constellations Aquarius (corresponding to Noah), Capricorn (corresponding to Ham), and the two Fishes of Pisces with the Great Square of Pegasus between them (corresponding to the brothers Shem and

Japheth walking with the "sheet" suspended between them – the Great Square corresponding to the sheet). In some artwork down through the centuries, a connection between Japheth and Sagittarius appears to be indicated as well.

An engraving showing the scene of Shem, Ham and Japheth with the drunken Noah, from 1493. The outline of Noah is strongly suggestive of the running figure of the constellation Aquarius in the sky. The outline of Japheth in this artist's conception actually corresponds to Sagittarius, and is positioned correctly for Sagittarius in relation to Ham (spelled "Cham" in the illustration), who corresponds to Capricorn. Note the left hand of Ham, which corresponds to the horn of the goat's head in the constellation.

Obviously, if Shem, Ham and Japheth actually correspond to constellations in the sky, then no people or ethnicity or culture on earth can be said to be *literally* descended from any of these three brothers, or from their father Noah who is Aquarius (a constellation which, for obvious reasons, plays a role in many mythological episodes involving water and floods).

On the other hand, we can *all* be said to act as "descendents of Ham" or as descendents of the other two brothers, when we either curse ourselves or our fellow men and women and seek to reduce them to a mere physical body and deny their divine nature, or when conversely we respect them and seek to remember their divine spiritual nature and to restore and uplift them, because they are more than just a physical body.

The terrible irony, of course, is that in turning the story of Shem, Ham and Japheth into a literalistic story about the supposed progenitors of different "races" of men and women on our planet, the literalizers are following the path of Ham in the story -- seeking to reduce their fellow men and women to their physical form, their material nature, and in doing so they upend and invert what I believe to be the entire message and meaning of the episode.

Thus, the restoration of the awareness that the ancient myths and scriptures of the world are speaking a celestial language has tremendous implications for human history -- both the very ancient and right up to our present day. The close correspondence of specific celestial references in myths and stories found in cultures separated by vast distances provides powerful additional evidence supporting the other evidence presented by researchers such as Graham Hancock and Robert Schoch for the existence of a lost civilization (or civilizations) in earth's very ancient past. And the incredible layers of meaning and wisdom contained in these ancient myths indicates that this lost civilization (or civilizations) possessed sophisticated spiritual advancement in addition to their other amazing capabilities.

The streams of that ancient wisdom which did in fact survive are still amazing and profound -- and constitute a precious inheritance which I believe was intended as a benefit for all men and women, who are all in a very real sense descendants of the survivors of whatever ancient catastrophe disrupted our remote enlightened ancestors.

The deliberate obstruction or even destruction of those remaining streams has been a terrible setback to humanity, with many tragic consequences. These consequences continue to be felt to this day, and shape the world and the societies in which we find ourselves at the present moment. How we choose to proceed from here will determine humanity's future course -- which is the subject explored in the next chapter.

Apollo pouring a libation, observed by a crow or a raven.

Humanity's Future History

In 1932, a book was published containing the account of the teachings of Oglala Lakota holy man Heháka Sápa -- Black Elk – recorded by Nebraska poet laureate John G. Neihardt, and entitled *Black Elk Speaks*.

In it, Black Elk says at one point:

> Once we were happy in our own country and we were seldom hungry, for then the two-leggeds and the four-leggeds lived together like relatives, and there was plenty for them and for us. But the Wasichus came, and they have made little islands for us and other little islands for the four-leggeds, and always these islands are becoming smaller, for around them surges the gnawing flood of the Wasichu, and it is dirty with lies and greed.[139]

There is a lot to examine in these two sentences. Black Elk chooses to characterize the difference between the two cultures by saying that his culture saw humanity as part of nature: they lived *together* with the earth's other creatures as *relatives*. In contrast, the representatives of the invading culture saw themselves as divided from nature, and in many ways hostile to nature, and that they created what Black Elk describes as "little islands" to physically separate people from one another and from the earth's other creatures.

In an article published after his visit to the protests taking place at Standing Rock (the very place where the great Hunkpapa Lakota holy man and leader Tatánke Íyotanke, Sitting Bull, was treacherously murdered) in opposition to an oil pipeline being built through sacred Native American sites and in violation of written requirements for any federal agency planning a project through land with which a Native nation may have cultural or historical connection to consult with the Native nations on a nation-to-nation basis, which did not take place in the routing of the Dakota Access Pipeline, Graham Hancock recounts the meaning of the term *Wasi'chu* as applied to the European settlers.

As Cody Two Bears, the District Representative of the Cannonball Community to the Standing Rock Sioux Tribal Council, explained to Graham Hancock, the word *Wasi'chu* means "he who takes the fat" or "he who takes the larger portion" -- in other words, a greedy person, a taker, an exploiter.[140]

This definition of the term -- and its application to the representatives of the invading culture which destroyed the Native cultures in North America -- fits with the concepts being articulated in the quotation from Black Elk cited above, because rushing to "take the fat" or to seize "the larger portion" comes from a mentality of scarcity which is implicit in the contrast that Black Elk is drawing.

The quotation from Black Elk implies two opposite views of humanity's place in the cosmos, and two opposite views of the cosmos itself: in the first, a vision of abundance -- as Black Elk says, "there was plenty for them and for us" -- and in the second, a vision of scarcity, described by Black Elk as a "gnawing flood [. . .] dirty with lies and greed."

I would argue that in those two sentences quoted from Black Elk Speaks, we can perceive the critical difference between the vision contained and preserved in the ancient wisdom of the world's myths, scriptures and sacred traditions and the inverted vision which resulted from the literalist twist that was imposed upon the ancient scriptures described in the preceding chapter, associated with the aggressive, intolerant, exclusivist reading of the scriptures in the Old and New Testament of the Bible by certain groups within the boundaries of the influence of the Roman Empire, which were carried from there to the rest of the planet.

One vision is a vision of plenty and a vision of connectedness between people and the wider world (both the wider material world and the Invisible World). The other vision is a vision of scarcity, and a vision of disconnection, disjointedness, and separation from both nature and the Invisible Realm which surges and pulses through the visible, material world at every point.

I would also argue that these two opposite visions can be directly linked to the differences between the two contrasting ways of understanding the ancient myths and sacred stories which we have been exploring in this book.

I am convinced, on the basis of overwhelming evidence (only a small sliver of which we have had space and time to explore in this particular volume) that the ancient myths are built upon a foundation of celestial metaphor -- and that they use the mighty cycles of the heavens, and the starry figures upon their infinite backdrop, as a powerful representation of the truth that this cosmos we inhabit is connected to and intertwined with and dependent upon an infinite and invisible dimension of pure potentiality.

This Other Realm -- the realm of the gods -- gives shape to the material realm that we perceive as ordinary reality. The relationship, according to the words of Black Elk cited earlier, is one of interdependence -- but the material realm proceeds from and thus is in a way subordinate to the infinite and invisible realm. As he says, "everything that we see here is something like a shadow from that world."[41]

This understanding of the material world as infused with, and interdependent upon, an invisible and infinite source-realm can be seen to inform virtually all of the ancient myths and traditional wisdom of the human race. We see it in the shamanic cultures. We see it in the myths of ancient Greece, and in the worldview, often cited, in which "each cave had its Faun or Dryad, – each fountain its Nymph."[42] We see it in the Vedas of ancient India, and in the worldview that infuses the Norse myths, and even in the writings of the Tao Te Ching.

I would argue that this vision of the cosmos follows directly from the understanding of the turning cycles of the heavens as representative of the endless interchange between the visible and the invisible, the finite and the infinite, the physical and the spiritual, which informs the world's ancient myths, scriptures, and sacred stories.

The contrary view – the vision of scarcity and disconnection – can likewise be seen to follow directly from the literalist, historicist, and exclusivist interpretation which insists upon seeing the stories like those of Shem, Ham and Japheth or of Jesus and his disciples as events which took place in a specific place and time, enacted by individuals who walked the earth thousands of years ago, and whose actions somehow divide humanity into different groups or subsets, such as the subsets of "those Jesus saved" and "those Jesus did not save," or "those descended from Shem," "those descended from Ham," and "those descended from Japheth."

By externalizing and historicizing stories which are actually celestial and metaphorical, we necessarily "externalize" and "physicalize" teachings which actually have more to do with internal and spiritual truths. This externalizing, by a very clear line of progression, can lead to disconnection, alienation, insecurity, and the greedy and grasping behavior encapsulated in the meaning of the word "Wasichu."

In the traditional worldview, the shamanic worldview, the vision of the cosmos in which the divine and infinite realm is present at all times and all places alongside and within this visible and material realm, there is a clear basis for understanding a connection between all human beings as well as a connection to all aspects of nature. The same invisible fabric connects them all. The same divine nature -- and indeed the same divine realm -- is present in them all.

But in the opposite vision, in which the stories are seen as historical accounts of events happening to actors in history long past (and the foundation of these episodes upon the motions of the heavenly cycles is denied as heretical), the connecting fabric is chopped up, and the divine nature is externalized, as it is in the literalistic interpretation of the figure of Jesus. The source of blessing must then be pursued, chased after, grasped at. It is given to some but by no means to all.

In his amazing book entitled *In the Dark Places of Wisdom*, Professor Peter Kingsley describes the alienation of chasing after externalities, even what we might call "spiritual or religious externalities." He writes:

> Even when we're finally where we want to be – with the person we love, with the things we struggled for – our eyes are still on the horizon. They're still on where to go next, what to do next, what we want the person we love to do and be. If we just stay where we are in the present moment, seeing what we're seeing and hearing what we're

hearing and forgetting everything else, we feel we're about to die; and our mind tortures us until we think of something else to live for. We have to keep finding a way away from where we are, into what we imagine is the future.

What's missing is more powerful than what's there in front of our eyes. We all know that. The only trouble is that the missingness is too hard to bear, so we invent things to miss in our desperation. They are all only temporary substitutes. The world fills us with substitute after substitute and tries to convince us that nothing is missing. But nothing has the power to fill the hollowness we feel inside, so we have to keep replacing and modifying the things we invent as our emptiness throws its shadow over our life.

[...]

Western culture is a past master at the art of substitution. It offers and never delivers because it can't. It has lost the power even to know what needs to be delivered, so it offers substitutes instead. What's most important is missing, and dazzling in its absence. And what we're offered is often just a substitute for something far finer that once used to exist, or still does exist, but has nothing in common with it except the name.

Even religion and spirituality and humanity's higher aspirations become wonderful substitutes. And that's what's happened to philosophy. What used to be ways to freedom for our ancestors became prisons and cages.[143]

This profound emptiness or "missingness" (as Peter Kingsley calls it) and the resulting desperate chasing after substitutes is familiar to us all.

As we've seen, the ancient wisdom contained in the myths provides us with the means of remedying this condition --

pictured in the restoring of "doubting Thomas" (the one who is never satisfied). The myths teach us that the infinite realm is always available -- and that it is the source of infinite potentiality, sufficient to satisfy our infinite missingness.

The solution, I'm convinced, must also involve the overcoming of the endlessly roving aspect of the mind depicted in figures such as Thomas or Psyche, and the "handing over of the reins" to Krishna as we see in the Bhagavad Gita (and as described in the Katha Upanishad). This solution involves meditation and connection with the Higher Self -- but this is not an external solution. The solution is available to us at any moment, right where we are. In the ancient myths, the gods appear instantly when they are evoked by name (we see this again and again in the Mahabharata or the Ramayana, or in the Norse myths when the name of Thor is spoken).

Clearly, although the ancient myths contain the remedy, there are many ways to miss it. As Peter Kingsley warns in the passage cited above, "Even religion and spirituality and humanity's higher aspirations become wonderful substitutes."

This warning can apply to any of the religions or spiritual systems built upon the ancient texts and ancient myths, including those built upon the Tao Te Ching or the Greek or Norse myths or any of the rest.

But, as Peter Kingsley notes, it is Western culture which above all is the "past master at the art of substitution." Could this be due to the fact that it was in the direct antecedent of Western culture -- the Western Roman Empire controlled by exclusivist literalist Christianity -- that the "heresy" of the inner divine nature was denied and prohibited most vehemently and most strictly, and the necessity of the acceptance of an *external* divine nature (in the person of a literal and historical Jesus) was demanded most insistently?

Stripped of the understanding of the cosmos as infused with and interpenetrated by the Other Realm at every point, declaring its implacable enmity to the shamanic worldview and to shamanic cultures in general, and decrying the vision of nymphs and dryads and satyrs and naiads protecting every fountain and glen as demonic, literalist Christianity has proven to be the most implacable foe of the gnosis that informs the ancient wisdom of the world's ancient myths and traditions. Little wonder, then, that the culture most thoroughly immersed in its teachings became the most incessant pursuer of external substitutes, the most insatiable consumer of material trinkets, the most aggressive colonizer of the world's resource-rich lands, and the source of the invaders who were given the name *Wasi'chu* by the Lakota, the "ones who grab the fat or the larger portion."

The two visions described by Black Elk could not be more stark in their contrast. One is a vision of plenty and of harmony, and one is a vision of scarcity and of divisiveness. Less than one hundred years after his words were published, the consequences of the second vision's ascendency could not be more evident. Western culture has tried to consume the entire world, and now appears to be in the process of trying to consume itself.

In many ways it seems that humanity finds itself at a crossroads between the visions described by Black Elk. Perhaps we always find ourselves at this crossroads, although many factors argue that the urgency of the choice between the two paths has never been greater than it is today.

The increased urgency derives from the forward rush of technology, certainly, but also from the rapidly increasing rate with which those most dedicated to the second vision (of disconnection and greed) are impoverishing the planet and imposing conditions of austerity, poverty, and oppression upon a larger and larger percentage of the world's population.

And once again I would argue that this dire situation can be directly traced back to the disconnection from the surviving streams of the ancient wisdom -- to the severing of the understanding of the interpenetration of this material realm at every point with the Invisible Realm, and to the understanding of the dependence of this material realm to that Invisible Realm . . . the realm of the gods.

At the beginning of this chapter is an image from an ancient kylix depicting Apollo in the act of pouring a libation of wine, while a crow or raven looks on. The outline of the seated god with extended arm evokes the outline of the constellation Virgo, which is very close to the constellation Corvus the Crow (although Corvus is actually located on the other side of Virgo from the side depicted by the ancient artist). Thus, the constellation of Corvus is closely associated with Virgo -- and ravens or crows are also associated with the spirit world, even as Apollo of course inhabits the realm of the gods, being a god himself.

The act of pouring a libation, described frequently in both the Iliad and the Odyssey, was done at the beginning of a meal, as a way of acknowledging that the food and drink which we enjoy, and upon which we depend for the sustenance of our lives, is itself a gift from the gods, and that it flows down to us from the spirit world.

The invisible spark of life which causes the grain to grow from a seed into the wheat from which we make our bread, or which causes the vines to grow which produce the grapes which will later be turned into wine, must come from the Invisible World. It is not a product of the material realm, and thus we are dependent upon the continued infusion of this physical world with the invisible, mysterious and divine outpouring which overflows from the spiritual world for the growth of the plants and vines and trees (and plankton) which sustain all life on earth.

The world's ancient wisdom acknowledges that these sources of life are gifts of the gods. In ancient Greece, the gifts of the vine and of grain and of the sea would be understood as having their ultimate source in the blessings bestowed by Dionysus and Demeter and Poseidon.

Conversely, we can fail to acknowledge this source of life. King Midas is a well-known figure from ancient Greek mythology famed for his bad judgment. He is most remembered for his request, when granted a single wish by the god Dionysus, to have everything he touched turn to gold -- a request which, when granted, made him so giddy with happiness that he could hardly believe his good fortune. As everyone knows, however, he soon came to regret that awful request.

There is another episode recorded by the ancients in which Midas again displays his bad judgment, this time when he was asked to judge a competition of musical skill between Apollo (the very god of music) and a satyr (in some accounts, a satyr named Marsyas, and in others the god Pan himself).

Apollo of course played upon a lyre, and the satyr upon the pan-pipes, and in some versions Midas, the King of Phrygia, was the sole judge of the contest, while in other accounts it was the mountain of Timolus itself (or the god of the mountain Timolus) which was to be the judge. In those versions, Timolus wisely judged that Apollo was the winner, but then Midas loudly disagreed with him and indicated that the satyr's playing had been more skilled, while in the versions in which Midas alone was the judge, he also unwisely selected the satyr as the winner of the contest.

As a punishment, Apollo gave Midas the ears of an ass, saying that the dull judgment of Midas and his lack of discernment in hearing should from then onwards be visible for all to see.

Both of these episodes have clear celestial foundations, examined below, and add to the already overwhelming body of evidence which supports the conclusion that virtually all of the myths are based upon the celestial system of metaphor, and thus have to do with imparting insights into the nature of the simultaneously material and spiritual universe in which we find ourselves, as well as into our own simultaneously material-spiritual nature as incarnate men and women.

And, as with other oicotypes examined throughout this volume, both of the episodes involving the terrible judgment of Midas find strong echoes in other well-known "judgment myths" involving very much the same theme, such as the famous "Judgment of Paris" which leads eventually to the Trojan War, and the equally-famous "Judgment of Solomon" which we have already encountered earlier in this volume and which is discussed in greater detail in *Star Myths of the Bible*.[144]

The episode we call the Judgment of Solomon, found in the Old Testament book of 1 Kings chapter 3 actually follows immediately after another "judgment" episode found in the very same chapter, in which Solomon in a dream or vision is visited by the Most High, who asks Solomon what he would like to be given.

This open-ended offer very much parallels the offer made to Midas in the myths of ancient Greece, in which Dionysus offers Midas one request. Midas unwisely asks for unlimited riches -- in the form of having anything he touches turn to pure gold. In contrast, in the vision of Solomon, the king asks for a wise and understanding heart, so that he can be a better ruler on behalf of the people. The text tells us that this request pleases the LORD, who says:

> 11 Because thou hast asked this thing, and hast not asked for thyself long life [literally, "many days"]; neither hast asked riches for thyself, nor hast asked the life of thine enemies; but hast asked for thyself understanding to discern judgment;
> 12 Behold, I have done according to thy words [. . .]
> 13 And I have also given thee that which thou hast not asked, both riches, and honor: so that there shall not be any among the kings like unto thee all thy days.[145]

Note that Solomon's request for wisdom and judgment is contrasted with other possible choices he might have made instead, such as riches, honor, long life, or power over his enemies.

In a similar manner, in the episode from Greek myth known as the Judgment of Paris, the youth named Paris (a prince of Troy or Ilium) is presented with a contest of beauty in which he must choose among three goddesses, each of whom offers him a reward if he will select her. The rewards offered include rulership and power (offered by Hera), heroism and fame (offered by Athena), and the hand of the most beautiful woman in the world to be his bride (offered by Aphrodite).

As we know, Paris awarded the contest to Aphrodite, and in doing so launched the Trojan War, because the most beautiful woman in the world, Helen, was already married to a king of the Achaeans, and all the other Achaean kings and heroes had (during the wooing of Helen) promised to defend whichever among them would be so fortunate as to win the right to marry her.

This disastrous decision by Paris again has parallels to the vision of Solomon, but Solomon decided not to request riches or honor but rather wisdom and judgment (both of which were conspicuously lacking in the characters of Paris and Midas). Solomon is told that because of this choice, he would also be given those things for which he did not ask, such as riches and honor.

Interestingly enough, the bad judgment of Midas brings to mind the masterpiece of esoteric fiction discussed earlier, the *Metamorphoses* of Apuleius (better known as "The Golden Tale of the Ass"), in which the narrator Lucian is transformed into a donkey, and undergoes a series of outrageous adventures before being restored to his original form by the goddess Isis herself.

Not only is the condition of Lucius in animal form reminiscent of Midas, but the restoration of Lucius comes very closely after a climactic episode in the story in which Lucius witnesses a re-enactment in a Roman arena of the mythical episode of the Judgment of Paris! In the discussion of the disastrous choice of Paris, Apuleius says through his narrator Lucian that Paris had "brought damnation upon mankind" by his desire to possess another's wife -- an interesting way of viewing the judgment and its consequences, especially because in the beauty contest judged by Paris, the choice of winner was indicated by the bestowing of an apple (can we think of any other ancient myths in which the selection of an "apple" resulted in the "damnation of mankind"? And note that the very word *apple* may have connections to the name of the god of music and the sun, *Apollo*).

All of these ancient judgment myths are built upon celestial metaphor. In the story of Midas and his golden touch (an example of extreme bad judgment), several ancient sources tell us that Midas was at first overjoyed at the granting of his request, but that he soon realized to his horror that he could neither eat nor drink anything without it also turning to gold (a situation that would soon end with his own death, unless his condition could be reversed).

In some versions of the story, the king's own daughter runs up to embrace her father and is herself transformed into solid gold. This particular aspect of the story does not seem to be present in many of the most ancient versions, but is perhaps the most well-known part of the Midas story today.

In almost every ancient version of the myth, Midas prays to the gods (usually to Dionysus, who originally granted Midas one request, but in other versions to Apollo) to have the curse of his golden touch taken away and reversed. Midas is told to go immerse himself at the source of the river Pactolus (whose source is found at the aforementioned Mount Timolus). In some versions, Midas is told to immerse his head three times in the stream. Thereupon, all the things (and people) turned to gold by Midas were restored to their original condition – and the river Pactolus from then on had golden sands which often yielded up gold flakes or gold nuggets.

For a variety of reasons, we can be quite confident that the story of King Midas is founded upon the region of the sky containing the constellation Perseus, who plays the role of the unwise king in this myth. Perseus is a constellation located near the arching path of the Milky Way galaxy, on the far side of the galactic band from the brightest and widest section between Sagittarius and Scorpio. Thus, the region at which Perseus stands beside and partly within the Milky Way can be envisioned as the "upper reaches" of the galactic river -- allegorized in this myth as the upper source of the river Pactolus. There, Perseus can be seen to be immersing himself in the river -- or even dunking his head in it!

The most dramatic part of the episode of Midas and his golden touch, of course, takes place when the his daughter runs to him to embrace him, and is herself turned to gold, to the king's horror. This aspect of the myth is almost certainly inspired by the outline of Perseus, and the outstretched arm on the western side of the constellation (the right side as we face the image of the star chart below). That side of the constellation Perseus reaches out and almost touches the constellation Andromeda, who plays the role of a beautiful maiden in many myths, and plays the role of the daughter of Midas in this story.

Can you see how the story of Midas touching his daughter can be clearly seen in the constellations? We can be very thankful that the god allowed Midas to change his mind and restore all that he had previously turned to gold, by immersing himself in the river!

I also believe that the very same constellations are involved in the story of the disastrous judgment of Midas at the music contest between the god Apollo and the satyr Marsyas (or, in some cases, Pan). This time, instead of playing the role of the beautiful daughter of the king, the constellation Andromeda actually plays the role of the satyr, with an arching tail and pan-pipes held aloft.

In this story, Apollo almost certainly corresponds to the constellation Sagittarius, which is an "archer" constellation associated with both Apollo and Artemis (who are brother and sister in the myths of ancient Greece). This correspondence is treated at greater length in *Star Myths of the World, Volume Two*.[46]

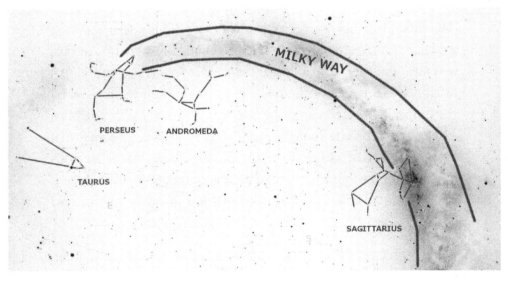

In the star chart, we can easily see how the portion of the Milky Way band near Perseus could be said to be the upper source of the celestial river itself – in fact, the galaxy is much fainter below Perseus and almost invisible, so that its source does seem to be around the head of the constellation Perseus. This is why the myth has Midas dunking his head three times at the river's source.

In the above diagram, we can also clearly see Midas reaching out to his daughter (who is played by the constellation Andromeda) and turning her to gold.

The same region also furnishes the celestial foundations for the music-contest. The bow of Apollo (played by the constellation Sagittarius) can also play the role of his famous lyre. And the ass-ears given to Midas are suggested by the proximity of the constellation Perseus to the outline of Taurus the Bull, directly below.

We know that the long horns of Taurus (seen in the chart above) were also seen as the long ears of an ass or donkey in some Star Myths. This is why the V-shaped Hyades (which form the two long sides of the triangular "head" of Taurus as outlined above) was described as the "jawbone of an ass" in the Samson story. Many mythical characters who are connected to the constellation Perseus ride upon the back of a donkey or an ass, such as Hephaestus in the myths of ancient Greece, and Balaam in the book of Judges in the Hebrew Scriptures (whose donkey eventually talks to him to ask Balaam why he keeps beating her -- and we can see that Perseus appears to be holding aloft a whip in one hand, furnishing some of the detail for that story).

Some readers may at first be skeptical of the claim that the same constellation who plays the role of the beautiful maiden in so many myths and who plays the role of the daughter of Midas in the golden touch episode could also play the role of the satyr in the music-contest story. Satyrs are famously animalistic in both their features and in their sexual appetites, and were usually depicted in surviving examples of ancient Greek pottery in the act of chasing after young women, often with their lascivious intentions very evident due to the fact that their sexual organs were also depicted in the artwork.

However, as the example below shows, satyrs in ancient Greek artwork were usually also depicted with a prominent and arching tail, similar to the tail of a fox, wolf, or even a horse. This tail corresponds to the upper limb of the constellation Andromeda which is usually envisioned as her upper leg, but which could also be imagined to be the tail of a satyr.

Note that the outline of Andromeda also features an upward-flung arm, from which extend outward two smaller lines marked by stars -- usually envisioned as the "chains" upon the wrist of the maiden when she is chained to the rock in the story of Perseus and the Gorgons. However, these "chains" in the sky could also be envisioned as the "pan-pipes" which satyrs in Greek myth are often fond of playing as their chosen musical instrument.

The illustration below shows an example of an ancient jar with black-figure artwork depicting a typical satyr with arching tail.

Note that the outline of Andromeda corresponds to the posture of the satyr (who is, in this case, the satyr named Silenus). The satyr's tail corresponds to the constellation's "upper leg," the satyr's rear leg corresponds to the constellation's "lower leg," and the satyr's forward leg corresponds to the "lower arm" of the constellation's outline. In this particular piece of art, no pan-pipes are depicted, but the Andromeda close-up shown below indicates their location in the stars:

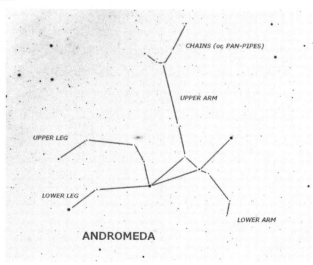

281

Having established that the Midas story corresponds to figures in the sky, we can then say definitively that the episodes in the Midas myth must have something to do with the interaction between the physical realm and the invisible realm.

The bad judgment of Midas involves an inability to rightly judge the proper relative value between the manifest realm upon which we focus most of our attention during this incarnate life and the realm of spirit, the realm of pure potentiality: the realm of the gods.

When he judges the music contest, Midas is depicted as making the outrageous judgment that the very god of music, the source in the realm of infinity for all musical talent and indeed for all music itself, could have been the loser in the contest. This decision represents a gross inversion of the natural order -- an elevation of the derivative above its source.

The judgment is made all the more egregious by having Midas award the prize to a satyr -- a being often representative of the animal or bestial nature, and of the dominance of gross physical drives over the higher spiritual aspirations.

When he is given one request from the god Dionysus, Midas (in contrast with Solomon) unwisely chooses gross physical wealth in the form of gold -- and soon realizes that such a single-minded focus on material riches has made him incapable of enjoying the far more essential blessings of food, and drink, and even familial love. The cold, dead metal is no substitute for the divine spark of life present in his daughter's beating heart -- and once he turns her to gold, he realizes that he has chosen unwisely in the extreme.

As we saw, the gods were benevolent enough to undo the rash request of Midas, as well as all the horrifying transformations he had wrought with his magic touch -- and they did so when Midas threw himself on their mercy, and followed their command to immerse himself in the river Pactolus at its heavenly source.

The failure to acknowledge the realm of spirit, from which all life and blessing in this material realm flows and has its source, ultimately debases and objectifies and deadens everything which should be a blessing and a sustenance and a source of joy and love. The failure to give primacy to the realm of spirit turns that which should be a blessing into a curse instead. Midas experienced this "first-hand" (we might say).

The reader of or listener to the story of foolish King Midas might be forgiven for concluding that no one could ever be foolish enough to make the disastrous mistakes that Midas made -- except that if we are honest, we will admit that we find *ourselves* following the exact same pattern of pursuing the "substitutes" that Peter Kingsley warned us about, and then wondering why they do not bring us any real and lasting satisfaction.

And in fact, if we can really absorb the lesson of the story of Midas, we might discover that our own society, in our own supposedly-enlightened modern age, has in fact rejected the profound lessons of the Midas myth, and now suffers from the very same horrendous consequences depicted in the ancient story of the king who failed to acknowledge the primacy of the gods and of the Invisible Realm, focusing only on gold and material riches.

We pollute the genes of corn and soybeans and zucchini and yellow squash and even salmon by genetically modifying their cells with the DNA of bacteria and with "triggers" derived from the genetic sequencing of viruses, as if the gifts of the earth are insufficient to support life. To paraphrase the arguments of Plutarch in his essay entitled "On the Eating of Flesh," this behavior "slanders the earth by implying she cannot support us, and impiously offends law-giving Demeter and brings shame upon Dionysus, lord of the cultivated vine, the gracious one, as if we did not receive enough from their hands."[47]

Those who support the creation of such monstrosities argue that without taking such irreversible measures, we would never be able to afford to feed all the people on earth -- an argument which obviously comes from the same mentality criticized by Black Elk in his prophetic warning from less than a hundred years ago, about the mentality of scarcity and of failure to live in harmony with nature. Because of this mentality, we are now feeding men, women and children with "scientifically-altered" plants (as well as animals fed with those plants) which not only contain altered genetic material and in some cases self-producing anti-bacterial properties which could prove very harmful to gut bacteria, but which are typically also drenched with far greater volumes of dangerous herbicides, because those crops have been genetically-modified to withstand more herbicide.

There are plenty of studies which suggest that these genetic modifications do not actually enable the feeding of more people at all -- and even if we suppose that they did, would the feeding of potentially very harmful food to more people be considered a step forward? How can feeding more people with something that could prove to be poisonous to their health (or which could even turn their own digestive systems into factories producing antibiotics) possibly be considered to be wise?

We may well be following in the footsteps of King Midas, who decided that he was not wealthy enough, and who then was horrified when he felt his food turning to cold, hard metal between his teeth, and his drink turning to metal as it splashed over his tongue and down his throat.

Similarly, we have been told by those following the vision of scarcity that we will never have enough electric energy unless we build nuclear reactors -- and it has now been over six years that massive volumes of radioactive wastewater from Fukushima has been pouring into the Pacific Ocean, with consequences that have yet to be fully understood but which could prove devastating to all life on the planet. Examples such as those offered above could be multiplied endlessly – and they either represent folly on the scale of King Midas, whose greed and inability to properly acknowledge the primacy of the life-giving gifts of the Invisible Realm (the real source of all life), or else they represent something even more diabolical: deliberate attacks upon life, by forces who have given themselves over to oppressing and enslaving and sickening and even exterminating (a possibility so foul that it is difficult to even contemplate for most normal men and women – but note that the world's myths do indicate the existence of malevolent forces in the Invisible Realm as well).

The world's ancient wisdom unanimously demonstrates that pursuing such a life-denying course -- whether from folly (as in the case of Midas) or from diabolical malevolence (as in the case of the head Kauravas in the Mahabharata) is self-destructive to those pursuing it, and never results in the outcome that those who pursue such a course either desire or anticipate.

But, as the story of King Midas also shows, even when things have seemingly spun irreversibly out of control, it is still possible to reverse course, acknowledge error, and turn to the mercy of the gods and beg for them to undo the disastrous consequences of bad judgment.

It may well be that the future of humanity depends upon such a course at this moment in history. Certain actors – by no means representing the majority of men and women on earth – have pursued the path of Midas to such a degree that our future generations (like the daughter in the Midas story) are threatened with their very survival, let alone their quality of life.

We may very well have already reached the point at which our future is turning to lifeless metal before our very eyes – as a result of our disconnection from and rejection of the ancient wisdom bequeathed to us in the world's incredible Star Myths.

Our only hope lies in acknowledging the error of our ways, an error that stems from an inability to rightly perceive our proper relation to the realm of spirit, the realm of the gods -- and, having acknowledged our error, to beg for mercy and demonstrate (as Midas did when he traveled to the heavenly source, and dunked his head) our dependence upon that realm.

Such a turning will involve, I believe, the acknowledgement that the divine realm of the gods is present in all men and women – in each and every one of them, regardless of their outward physical appearance or form or place of origin -- and the acknowledgement that all of the world's ancient myths and scriptures are closely related, and none has primacy or right to exclusivity over any of the others (and that in fact, all of them are valuable sources of clues to help us piece together the picture that today exists in fragmented form, with gaps and "misty" spots, like the mist-painting described in *Hamlet's Mill*).

And it will also involve a serous attempt to reject the vision of scarcity and disconnection that Black Elk linked to the term *Wasi'chu* -- the ones who take the fat or the bigger portion -- and to replace it with a vision of plenty, and of connection with the rest of the cosmos: with a conception of Nature that includes the Invisible Realm as well (present in all parts of Nature at every moment and in every point).

To exhibit such a vision on a societal level -- a vision of plenty and of connection and integration with others and with the natural and spiritual cosmos -- will require us to pursue the practices that will encourage our own inculcation of such a vision on an individual level as well, one that heeds the warnings given by teachers such as Black Elk and Peter Kingsley, by following the disciplines given for our help in this department, such as Yoga or meditation or the practice of stillness. In this pursuit, we have access to tremendous ancient wisdom in the world's treasury of myths, including examples we have explored in this book such as the story of Eros and Psyche, of Jesus and Thomas, and of Krishna and Arjuna.

But such individual integration with the Invisible and Infinite can help others and society at large only to the degree that it leads to changes in the vision that informs the direction of policies and large-scale actions.

A society guided by the first vision described by Black Elk, trusting in plenty and in the connection of all beings, will exhibit radically different policies towards providing for those who are aged or sick or infirm or hungry or homeless. And it will exhibit a radically different attitude towards seizing the resources of others, or instigating wars and so-called "revolutions" to give foreign corporations access to their mineral or agricultural resources or public infrastructure.

And, a populace animated by the positive vision described by Black Elk will have a very different response when arguments are presented that we must accept genetically-modified food or nuclear power-plants in order to "have enough" (both of which are extremely questionable arguments animated by a "vision of scarcity" which is scarcely less short-sighted than that exhibited by King Midas).

Perhaps most important, however, is the realization that even if the vast majority of the people of the world do not desire wars, or genetically-modified foods, or nuclear power-plants generating billions of years' worth of radioactive waste (stored on-site in most cases), the power structures that resulted from the dissolution of the Roman Empire after its takeover by the same forces of exclusivist Christian literalism which decisively severed the links to the ancient wisdom in what is today known as "the West" are still in place, and these power structures still enable well-placed and well-connected members of small groups of men and women to steer the world into wars and into adoption of genetically-modified food and all the rest, against the wishes and against the best interest of the majority of the world's population, for their own purposes.

The hallmark of the literalist Christian takeover in antiquity was an attitude that "we are right and everyone else is wrong" – that some scriptures or teachings are better than others, justifying the suppression or destruction of "the others." This attitude has been the hallmark of "the West" (or at least the attitude of many of its leaders) ever since, not only towards other sacred traditions or other gods, but also towards entire cultures and classes and regions of the world.

It is quite possible that the same families that pulled off this literalist Christian takeover during the second through fifth centuries AD within the Roman Empire became the so-called "nobility" of the Middle Ages who carved up the ownership of the land from the western shores of Europe and the British Isles to the steppes of Russia, and whose families also furnished the men who would hold positions of power within the literalist Christian hierarchy for centuries to come.

These conspirators against the gods gained control of tremendous wealth and power, and the shape of society to this day still reflects the imprint of structures erected during the centuries of feudal control following the dissolution of the Roman Empire. Attempts to re-shape this system or to escape from it have proceeded in fits and starts, but despite these efforts many aspects of those structures are still in place, and new manifestations of the old feudalism (in some ways even more virulent than the older version) have been implemented to undo the work of those who sought to dismantle the feudal system.

On April 4, 1967, Martin Luther King, Jr. gave his speech *Beyond Vietnam*, in which he addressed the political and military actions of his country in the war in Vietnam -- but in terms of issues that go far beyond that specific war.

In that speech, he declared that "The war in Vietnam is but a symptom of a far deeper malady within the American spirit" -- and he could have also expanded that to include "the West," since the situation in Vietnam was certainly also the product of European imperialism before the United States became involved.[48]

Martin Luther King then went on to say that:

> In 1957 a sensitive American official overseas said that it seemed to him that our nation was on the wrong side of a world revolution. [. . .] I am convinced that if we are to get on to the right side of the world revolution, we as a nation must undergo a radical revolution of values. We must rapidly begin -- we must rapidly begin the shift from a thing-oriented society to a person-oriented society. When machines and computers, profit motives and property rights, are considered more important than people, the giant triplets of racism, extreme materialism, and militarism are incapable of being conquered.[49]

This struggle which Dr. King called a "world revolution" was linked to the struggle against those very structures of imperialism that stretch all the way back through history to the severance of the connection to the ancient wisdom by the families who later carved up the Western Empire and later became colonial powers that continue to operate to this day.

According to Dr. King, as well as to the official he mentions in this part of his speech, those directing policy in the United States were clearly taking sides in that struggle, and on the wrong side. To get back on the right side, Dr. King declared, would require "the shift from a thing-oriented society to a person oriented society" -- the very lesson that Midas had to learn before he realized that he himself had chosen the wrong side.

The choice by Midas of "things" instead of "life" threatened to destroy of everything that he loved (in the pursuit of wealth) and threatened his own imminent death by starvation or thirst unless the gods could be persuaded to show him mercy.

I believe that it is very possible that the very same forces which declared the "pagan myths" to be invalid and devoid of worth are the same ones who have been valuing "profit motives and property rights over people" for many centuries, and who have shown themselves willing to start wars based on that exact formulation of relative valuation.

Because of this reality, a change from the self-destructive path of Midas, and from the *Wasi'chu* vision described by Black Elk, will require a very large number of people to wake up to the danger and "return to the source of the Pactolus" (so to speak) -- including some of those who have access to the levers of power. The vision of scarcity is very powerful -- but it is in fact false, and the Infinite Realm (the source-realm and the seed-realm) can in fact provide the remedy to the folly of Midas.

But, as recent history illustrates, the forces arrayed on the wrong side of that struggle will not change easily. One year to the day after giving his powerful speech *Beyond Vietnam*, Martin Luther King, Jr. was foully and treacherously murdered by well-organized and well-connected forces valuing profit motives and property rights over people, and serving the "giant triplets of racism, extreme materialism, and militarism" that Dr. King had exposed in his speech.

The fact that these forces felt the need to murder Martin Luther King shows that his message -- and the possibility that millions of people would recognize the truth of his words and would change their own vision because of his words -- constitute a real threat to those giant triplets, and to the power structures that seem so intractable and insurmountable, but which can in fact be overcome.

It is also undeniable that much of the power of Dr. King's message came from his personal moral clarity, rooted in his own spirituality, his familiarity with the Biblical scriptures -- and his deep apprehension of their message of the worth and value and sanctity of all men and women regardless of their race or country or external circumstances.

The gods could solve even the case of a King Midas. It is a question of vision -- and the folly of Midas was rooted in his inability to properly acknowledge and value the gifts of the gods. But when his vision was corrected, they remedied the disasters his flawed vision had wrought.

And, perhaps the most hopeful point to remember is that the gifts of the gods work themselves into the world most potently through men and women themselves, according to the teachings we find in the world's ancient wisdom. The gods and goddesses appear at a moment's notice when evoked because they are always present, and always available.

In Mahabharata Book 6 (Bhishma Parva) and section 23, it is the very eve of the battle of Kurukshetra, and the penultimate section of the epic prior to the sections which make up the Bhagavad Gita. Krishna and Arjuna are surveying the endless ranks of the enemy forces, drawn up in battle array, and

Krishna urges Arjuna to utter his hymn to the goddess Durga and request her aid in the upcoming struggle.

Arjuna does so, addressing her as the one who is identical with Brahman, the one who is endowed with every auspicious attribute, the one who dwells in accessible regions, the one who is the beauty of all creatures, and the one who supports both Sun and Moon and causes them to shine.

Perceiving Arjuna's devotion, Durga appears at the conclusion of his hymn. She tells Arjuna that within a short time he shall conquer his foes, and that he is in fact "incapable of being defeated by foes, even by the wielder of the thunderbolt himself."[150]

Alvin Boyd Kuhn declares (and backs it up with volumes of evidence) that the ancient wisdom teaches "the entangling of deity with mortal flesh in all humanity."[151] The divine powers of the Invisible Realm are working their way out through all men and women at all times, and this is too much for anyone (no matter how well-connected, well-organized, or well-armed) to keep down indefinitely.

For far too long, we have been divorced from the ancient wisdom. We have had the myths and the ancient texts in our midst, and yet we have not always known what language they are speaking. We have listened to what those who have offered to translate their words have said to us, without realizing that they might have been either misunderstanding the language being spoken, or else deliberately mis-translating the myths and misleading us about what these ancient visitors are trying to relate.

The world's ancient wisdom is a precious inheritance given to all men and women, to put us in touch with the very realm of Infinity, and with the gods and goddesses who occupy the Invisible Realm, and to guide us towards integration with our Higher Self and elevation of our own inner divine nature.

They are speaking to us in a language of celestial metaphor, and of the heavenly cycles. And their purpose is to be a real and potent source of blessing to us, as we negotiate this battlefield of Kurukshetra.

> *The goddess said,*
> *"Within a short time thou shalt conquer thy foes, O Pandava.*
> *O invincible one, thou hast Narayana again for aiding thee.*
> *Thou art incapable of being defeated by foes, even by the wielder of the thunderbolt himself."*
> *Having said this, the boon-giving goddess disappeared soon.*
> *There where righteousness is, there is Krishna;*
> *and there where Krishna is, there is victory.*[152]

Durga

End Notes

1. Dundes, *International Folkloristics*, 138.
2. Taylor, *Devil's Pulpit*, 65.
3. Taylor, *Astronomico-Theological Lectures*, 357.
4. *Ibid*, vi.
5. *Devil's Pulpit*, x.
6. Rolleston, *Mazzaroth*, page 3, table B2.
7. Seiss, *Gospel in the Stars*, 37.
8. *Ibid*, 14.
9. De Santillana and von Dechend, *Hamlet's Mill*, 424, 216, 300, 328.
10. *Ibid*, 166 for Samson and 424 for Revelation 9.
11. *Ibid*, 330.
12. *Ibid*, 329.
13. *Ibid*, 328.
14. *Ibid*, 5. Note also that the final two lines of this famous paragraph are an adaptation of two lines from a poem by William Morris (1834 - 1896) entitled "The Argonauts and the Sirens," from the text *The Life and Death of Jason* (1867), in which Morris writes: "Their words are no more heard aright / Through lapse of many ages."
15. Rey, *The Stars: A New Way to See Them*, 10.
16. *Ibid*.
17. *Ibid*.
18. Mathisen, *Star Myths of the World, Volume One*, 65.
19. *Ibid*, 106.
20. Luomala, *Maui of a Thousand Tricks*, 70, 72, 138 - 226.
21. See for example discussions in Kuhn, *Lost Light*, 462 - 468.
22. *Ibid*, 10.
23. *Ibid*, 154.
24. *Ibid*, 517.
25. Kuhn, *Easter: the Birthday of the Gods*, 8 - 11. Note that the spelling of "fulness" is here rendered as it appears in the original.
26. As argued by Walter Cruttenden with extensive supporting evidence in *Lost Star of Myth and Time* – and more supporting evidence continues to be discovered which appears to support his hypothesis.
27. See Schwaller de Lubicz, *Esoterism & Symbol*, 1, 3, 75, and also Mathisen, *Undying Stars*, 17 - 25.

28. See *Hamlet's Mill*, 423 - 424, and *Undying Stars*, 9 - 13, and *Star Myths of the World, Volume Three*, 714 - 716.
29. *Star Myths of the World, Volume One*, 422 - 430.
30. *Lost Light*, 416.
31. *Star Myths of the World, Volume Three*, 54 - 58.
32. Neugebauer, *Exact Sciences in Antiquity*, 83 - 87.
33. Neugebauer and Parker, *Egyptian Astronomical Texts*, Volume 1, page 24, footnote 1. See also Sellers, *Death of Gods in Ancient Egypt*, 24 - 25.
34. Sellers, 196.
35. Sellers, 84 - 108.
36. See Mathisen, *Mathisen Corollary* 97 - 98, 126 - 134, *Undying Stars* 211 - 216, and *Star Myths of the World, Volume One*, 327 - 334.
37. See *Star Myths of the World, Volume Two*, 246 - 247, 644 - 646.
38. *Ibid*, 626 - 633.
39. Buckley, Odyssey, 200 - 202.
40. *Star Myths of the World, Volume Three*, 606 - 609.
41. Luomala, 43.
42. *Ibid*, 39.
43. *Ibid*, 144.
44. *Ibid*.
45. Plutarch, *De Iside et Osiride*, 15 - 16.
46. *Star Myths of the World, Volume One*, 386 - 392.
47. *Star Myths of the World, Volume Three*, 378.
48. *Hamlet*, III. i.
49. See Iyengar, *Light on Yoga*, 23 - 36.
50. Kojiki I. 4. 6 - 12. Phillips trans., 50 - 51.
51. *Star Myths of the World, Volume Three*, 263 - 265.
52. George, introduction to *Epic of Gilgamesh*, 52.
53. *Hamlet's Mill*, 218. Italics in original.
54. George, 88.
55. *Ibid*, 95.
56. Edda, 61.
57. Apollodorus, *The Library*. Frazer trans. I.5.1. pages 35 - 37.
58. Diodorus, *Library of History*. Oldfather trans, V. 4.
59. *Ibid*.
60. *Star Myths of the World, Volume One*, 438 - 454.
61. Kojiki XVI. Chamberlain, trans., 63 - 65.
62. Cosmopoulos, *Bronze Age Eleusis*, 19.

63. Kojiki VII. Chamberlain trans., 33.
64. Kojiki IX. Chamberlain trans., 39 - 40.
65. Gayton, "Orpheus Myth in North America," 266.
66. *Ibid*, 263 - 264.
67. *Ibid*, 267.
68. *Ibid*, 271.
69. De Brébeuf, *Jesuit Relations*, vol 10, chapter 2, page 147.
70. *Ibid*, 149 - 151.
71. Gayton, 283.
72. *Ibid*.
73. Hultkrantz, *North American Indian Orpheus Tradition*, 149.
74. Edda, 50.
75. *Ibid*, 51.
76. Gayton, 285.
77. Hultkrantz, 307.
78. Neihardt, *Black Elk Speaks*, 67.
79. Pindar, *Odes*, Svarlein trans.
80. *Devil's Pulpit*, 102.
81. *Ibid*, 102 - 103.
82. *Undying Stars*, 326 - 327.
83. *Star Myths of the World, Volume Three*, 722 - 732.
84. Book of Thomas the Contender, Turner trans., page 138, line 8.
85. And note also the important discussion in *The Cult of the Heavenly Twins*, by J. Rendel Harris, regarding the tradition of linking Judas (described as the "brother of Jesus" in Matthew 13: 55) with Thomas. The discussion is found in *Cult of the Heavenly Twins* especially on pages 117 - 119. On page 117, Harris writes: "We knew already that Thomas to the Syrians was Judas Thomas; that explanation is in the Syrian New Testament as well as in the Apocryphal Acts." He goes on to say, "It is not generally known that there are other Apostles suggested by traditions as twins or even as twins of Jesus, and apparently with the same object of explaining the meaning of Thomas in the apostolic lists: thus we have something very like a case of Jacobus Thomas to set over against the Syriac Judas Thomas" (117) and later Harris says, "The Edessan belief in the twinship of Jesus and Judas Thomas must, in any case, belong to the earliest period of Christianity in that city" (118).
86. Harris, 58ff.
87. *Ibid*, 58 - 62.

88. *Ibid*, 46. Harris is here citing the ancient author Pausanias who describes the Chest of Cypselus in Book 5 of his *Description of Greece*. Specifically, Pausanias describes this chest in his section on Ellis, beginning in chapter 18 and section 5 and continuing through chapter 19 and section 10. The description of the Dioscuri which Harris is referencing is found in chapter 19 and section 1.

89. *Ibid*, footnote on page 23. Harris is here referencing a passage in the last order of the Mishnah in the Talmudic literature, which is the order or divisions (*sedarim*) entitled Tohorot. The discussion of the twins which Harris is referencing is found in Niddah 27.

90. See for instance *Lost Light*, 370 - 371.

91. Harris, 46. See also *Undying Stars*, 206 - 217, regarding the mysteria of the Great Gods of Samothrace.

92. Harris, 46.

93. *Lost Light*, 130, 171.

94. *Ibid*, 130 - 131.

95. Katha Upanishad, Part I, chapter 1, verse 22 - chapter 2, verse 3.

96. *Ibid*, Part I, chapter 2, verse 7 through chapter 3, verse 8.

97. Iyengar, *Light on Yoga*, 30.

98. Proverbs 18: 24.

99. Apuleius, 122.

100. Odyssey, Buckley trans., pages 108 - 109.

101. *Ibid*, 109.

102. *Star Myths of the World, Volume One*, 376.

103. See also *ndying Stars*, 39.

104. Blofeld, *Mantras*, 6 - 7.

105. Yoga Sutras, I. 27 - 29. Steiner, trans. Explanatory material shown inside brackets is found in the original source cited here in the Bibliography.

106. I learned this perspective on the sacred word ॐ from Boris Fritz. Conversation between the author and Boris Fritz at the 10th *Conference on Precession and Ancient Knowledge* in Rancho Mirage, California, 10/01/2016.

107. Iyengar, 50.

108. Temple, *Sirius Mystery*, 266.

109. Waters, *Book of the Hopi*, 3.

110. Temple, 130 - 131.

111. Kuhn, *Esoteric Structure of the Alphabet*, 20. See also *Star Myths of the World, Volume Three*, 449 and also *Undying Stars*, 101 - 102.

112. *Star Myths of the World, Volume Three*, 378 - 384.

113. Bonacci, *Your Body is the Holy Land, part I* (video), 0:25:00 and following.
114. Bonacci, *Jacob and the Pineal, Horus the Newborn Sun & Crossing the Red Sea* (video), 0:06:30 and following.
115. *Huang Di Nei Jing Ling Shu*, Unschuld trans., 4.
116. *Huang Di Nei Jing su wen*, Unschuld, Tessenow, and Zheng trans., 56.
117. *Ibid*, 45 - 49.
118. *Ibid*, 91 - 93.
119. *Ibid*.
120. *Ibid*, 176.
121. *Ling Shu*, Wu trans., 7.
122. *Ibid*, 20.
123. Chapter 1, section 3.
124. Matthew 12: 43 - 45 and see also Luke 11: 24 - 26.
125. List of ancient texts dealing with sex comes from Van Gulik, *Sexual Life in Ancient China*, 121- 122; suggestion that incidents in ancient scriptures such as the Onan encounter might have something to do with esoteric teachings on prevention of ejaculation and related "internal alchemy" topics is my own speculation.
126. *Ibid*, 123.
127. See West, *Serpent in the Sky*; Hancock, *Fingerprints of the Gods*; Hancock and Faiia, *Heaven's Mirror*; Schoch, *Forgotten Civilization*; and Alison, *Prehistoric Alignment of World Wonders*.
128. *Forgotten Civilization*, 37.
129. *Ibid*, 40.
130. *Ibid*.
131. *Ibid*.
132. *Ibid*, 42.
133. *Ibid*, 53 - 57.
134. *Ibid*, 54 - 57.
135. Hancock, *Magicians of the Gods*, 308 - 325.
136. *Ibid*.
137. *Ibid*, 318 - 319.
137. *Ibid*, 327 - 333.
138. *Hamlet's Mill*, 4.
139. *Black Elk Speaks*, 9.
140. Hancock, "Standing Rock: Water is Life."
141. *Black Elk Speaks*, 67.

142. Trant, *Narrative of a Journey Through Greece*, 133. This volume is probably not the original source of this phrasing of "each fountain had its nymph" -- an expression which is found in dozens of books from the nineteenth century. Trant's is one of the oldest I could definitively locate, but there may have been eighteenth century or earlier expressions of the same idea.

143. Kingsley, *In the Dark Places of Wisdom*, 33 - 36.

144. *Star Myths of the World, Volume Three*, 544.

145. I Kings 3: 11 - 13.

146. *Star Myths of the World, Volume Two*, 254 - 259, 348 - 361.

147. Plutarch, *De esu carnium* ("On the Eating of Flesh").

148. Martin Luther King, Jr., *Beyond Vietnam*.

149. *Ibid.*

150. Mahabharata 6.23, Ganguli trans.

151. *Lost Light*, 128.

152. Mahabharata 6.23, Ganguli trans.

Illustration Credits

listed by page number

21. Virgo illustration by Sidney Hall (1825). Wikimedia commons.
https://commons.wikimedia.org/wiki/File:Sidney_Hall,_Virgo,_1825.jpg

25. The Pythia and Aegeus. Wikimedia commons.
https://commons.wikimedia.org/wiki/File:Themis_Aigeus_Antikensammlung_Berlin_F2538_n2.jpg

29. Sagittarius illustration by Sidney Hall (1825). Wikimedia commons.
https://commons.wikimedia.org/wiki/File:Sidney_Hall_-_Urania%27s_Mirror_-_Sagittarius_and_Corona_Australis,_Microscopium,_and_Telescopium.png

33. Artemis slaying Actaeon. Name-vase of the Pan Painter. From John Davidson Beazley. *Attic Red-figured Vases in American Museums.* Cambridge: Harvard UP, 1918. page 113.
https://books.google.com/books

35. Odysseus slaying the suitors. Wikimedia commons.
https://commons.wikimedia.org/wiki/File:Page_229_fig_16,_inset_illustration._Folk-Lore,_vol._14.png

36. Hercules illustration by Sidney Hall (1825). Wikimedia commons.
https://commons.wikimedia.org/wiki/File:Sidney_Hall_-_Urania%27s_Mirror_-_Hercules_and_Corona_Borealis.png

39. Heracles and the Amazons. Wikimedia commons.
https://commons.wikimedia.org/wiki/File:Heracles_Amazons_Met_61.11.16.jpg

40. Gilgamesh or Enkidu wrestles a Lion. From the frontispiece of *Chaldean Account of Genesis*, by George Smith, where it is identified as being from a Babylonian cylinder-seal. NY: Scribner, Armstrong & Co., 1876.
https://archive.org/stream/chaldeanaccountooosmit#page/n3/mode/2up

42. Hanuman sculpture. Wikimedia commons.
https://commons.wikimedia.org/wiki/File:Bajrang_Hanuman_ShrutikaMurthy.JPG

59. Zodiac wheel (used throughout in multiple illustrations, with modifications). Wikimedia commons.
https://commons.wikimedia.org/wiki/File:Astrological_signs_by_J._D._Mylius.jpg

77. Isis receiving the Djed pillar. Wikimedia commons.
https://commons.wikimedia.org/wiki/File:Abydos_seti_16_det2.JPG

74. Star-calendar from tomb of King Seti I. Wikimedia commons.
https://commons.wikimedia.org/wiki/File:Seti1-Lepsius-III-137-b.jpg

83. Star-calendar from the tomb of Senenmut.
https://commons.wikimedia.org/wiki/File:Senenmut.jpg

84. Osiris with Isis in hawk- or falcon-form. From Wallis Budge, *Osiris and the Egyptian Resurrection, Volume 2*, page 42.
https://archive.org/stream/osirisegyptianre02budg#page/n7/mode/2up

114. Judgment of Solomon. Peter Paul Rubens. Wikimedia commons.
https://commons.wikimedia.org/wiki/File:Peter_Paul_Rubens-The_Judgement_of_Solomon-Statens_Museum_for_Kunst.jpg

115. Wisdom of Solomon. James Jacques Joseph Tissot. Wikimedia commons.
https://commons.wikimedia.org/wiki/File:Tissot_The_Wisdom_of_Solomon.jpg

141. The Finding of Moses, Gustave Doré. Wikimedia commons.
https://commons.wikimedia.org/wiki/File:031.The_Finding_of_Moses.jpg

154. Daikoku and Otafuku. Wikimedia commons.
https://commons.wikimedia.org/wiki/File:Gyosai_Daikoku.jpg

200. L'Incredulità di San Tommaso, Giovanni del Giglio. Wikimedia commons.
https://commons.wikimedia.org/wiki/File:L%27incredulita_di_San_Tommaso.jpg

206. Eros (or Cupid) and Psyche, Charles Paul Landon. Wikimedia commons.
https://commons.wikimedia.org/wiki/File:Annales_du_musée_et_de_l%27école_moderne_des_beaux-arts_-_recueil_de_gravures_au_trait,_d%27après_les_principaux_ouvrages_de_peinture,_sculpture,_ou_projets_d%27architecture,_qui,_chaque_année,_ont_remporté_(14762994904).jpg

234. Vision of Jacob at Bethel. Wikimedia commons.
https://commons.wikimedia.org/wiki/File:Schnorr_von_Carolsfeld_Bibel_in_Bildern_1860_032.png

259. Shem, Ham and Japheth with Noah, 1493. Wikimedia commons.
https://commons.wikimedia.org/wiki/File:Nuremberg_chronicles_-_f_15v.png

262. Apollo pouring a libation. From a tomb at Delphi. Wikimedia commons.
https://commons.wikimedia.org/wiki/File:Apollo_black_bird_AM_Delphi_8140.jpg

281. Silenus. Wikimedia commons.
https://commons.wikimedia.org/wiki/File:Silenos_Midas_Met_49.11.1.jpg

295. Durga. Temple of Durga at Aihole, India. Wikimedia commons.
https://commons.wikimedia.org/wiki/File:Le_temple_de_Durga_(Aihole,_Inde)_(14196864577).jpg

Illustrations not listed above were created by the author. Star charts created using Stellarium (stellarium.org).

Bibliography

Alison, Jim. *Prehistoric Alignment of World Wonders: A New Look at an Old Design.* www.home.hiwaay.net/%7Ejalison/index.html

Apollodorus. *Apollodorus: The Library.* Trans. James George Frazier. 2 vols. London: Heinemann, 1921. www.perseus.tufts.edu

Apuleius. *The Golden Ass.* Trans. Jack Lindsey. Bloomington, Indiana: Indiana UP, 1960.

Blofeld, John. *Mantras: Sacred Words of Power.* London: Allen & Unwin, 1977.

Bonacci, Santos. *Jacob and the Pineal, Horus the Newborn Sun & Crossing the Red Sea* (video). YouTube video published on 07/01/2016 by Mr. Astrotheology.
https://www.youtube.com/watch?v=elm41W-ptNM

Bonacci, Santos. *Your Body is the Holy Land, part 1* (video). YouTube video published on 06/04/2012 by Mr. Astrotheology.
https://www.youtube.com/watch?v=EvrG4oqXjM

The Book of Thomas (the Contender). Trans. John D. Turner. Gnostic Society Library. www.gnosis.org

Burley, Paul. "Göbekli Tepe: Temples Communicating an Ancient Cosmic Geography." Originally written in June, 2011 and published in March, 2013 at:
www.grahamhancock.com/forum/BurleyP1.php.

de Brébeuf, Jean. *Jesuit Relations and Allied Documents: Travels and Explorations of the Jesuit Missionaries in New France, 1610 - 1791.* 71 vols. Trans. Reuben Gold Thwaites. Cleveland: Burrows Brothers, 1898. Volume 10.
moses.creighton.edu/kripke/jesuitrelations/relations_10.html

Cosmopoulos, Michael B. *Bronze Age Eleusis and the Origins of the Eleusinian Mysteries.* NY: Cambridge UP, 2015.

Cruttenden, Walter. *Lost Star of Myth and Time.* Pittsburgh: St. Lynn's Press, 2005.

De Santillana, Giorgio and Hertha von Dechend. *Hamlet's Mill: An essay on myth and the frame of time.* Boston: David R. Godine, 1969. First paperback ed. 1977.

Diodorus Siculus. *Library of History.* Trans. Charles Henry Oldfather, C. L. Sherman, C. Bradford Welles, Russel M. Geer, and F. R. Walton. Loeb Classical Library. 12 vols.
www.penelope.uchicago.edu

Dupuis, Charles François. *Origin of All Religious Worship: Translated from the French of Dupuis.* New Orleans: Müller, 1872.
Originally published in France as *l'Origine de tous les Cultes*, 1781.
https://archive.org/stram/originallreligioodupugoog#page/no/mode/2up

Epic of Gilgamesh: The Babylonian Epic Poem and Other Texts in Akkadian and Sumerian. Trans. Andrew George. NY: Penguin, 1999.

Gayton, Anna H. "The Orpheus Myth in North America." *Journal of American Folk-Lore*, Volume 48, Number 189 (July to September, 1935). Pages 263 - 293.

Hall, Sidney. *Urania's Mirror*, a set of illustrated cards accompanying Jehoshaphat Aspin's *A Familiar Treatise on Astronomy.* London: Samuel Leigh, 1825.

Hancock, Graham. *Fingerprints of the Gods.* NY: Crown, 1995.

Hancock, Graham and Santha Faiia. *Heaven's Mirror: Quest for the Lost Civilization.* NY: Three Rivers, 1998.

Hancock, Graham. *Magicians of the Gods: The Forgotten Wisdom of Earth's Lost Civilization.* NY: Thomas Dunne, 2015.

Hancock, Graham. "Standing Rock: Water is Life." Published 12/24/2016. www.grahamhancock.com/hancockg16-standingrock/

Harris, J. Rendel. *The Cult of the Heavenly Twins.* Cambridge: Cambridge UP, 1906.

Higgins, Godfrey. *Anacalypsis: an Attempt to Draw Aside the Veil of the Saitic Isis, or an Inquiry into the Origin of Languages, Nations and Religions*. London: Longman, Rees, Orme, Brown, Green, and Longman, 1836.

Homer. *Odyssey*. Theodore Alois Buckley, trans. London: Henry G. Bohn, 1851. https://books.google.ca/books

Huang Di nei jing su wen: An Annotated Translation of Huang Di's Inner Classic -- Basic Questions. Trans. Paul U. Unschuld, Hermann Tessenow, and Zheng Jinsheng. Berkeley: UCalifornia Press, 2011.

Huang Di Nei Jing Ling Shu: The Ancient Classic on Needle Therapy. Trans. Paul U. Unschuld. Berkeley: UCalifornia Press, 2016.

Hultkrantz, Åke. *The North American Indian Orpheus Tradition: A Contribution to Comparative Religion*. Stockholm: Ethnographical Museum of Sweden, 1957.

International Folkloristics: Classic Contributions by the Founders of Folklore. Alan Dundes, ed. Lanham, Maryland: Rowman & Littlefield, 1999.

Iyengar, *Light on Yoga*. Revised ed. NY: Schocken Books, 1979.

Katha Upanishad. Trans. Swami Nikhilananda.
www.gayathrimanthra.com

King, Martin Luther, Jr. *Beyond Vietnam* (speech). 04/04/1967.

Kingsley, Peter. *In the Dark Places of Wisdom*. Point Reyes, California: Golden Sufi Center, 1999.

Kojiki. Donald L. Philippi, trans. Tokyo: Tokyo UP, 1968.

Kojiki. Basil Hall Chamberlain, trans. Tokyo: Asiatic Society of Japan, 1919.
www.sacred-texts.com

Kuhn, Alvin Boyd. *Easter: The Birthday of the Gods*. Second Ed. Wheaton, Illinois: Theosophical Press, 1966.

Kuhn, Alvin Boyd. *Lost Light: An Interpretation of Ancient Scriptures.* Elizabeth, NJ: Academy Press, 1940.

Kuhn. *The Esoteric Structure of the Alphabet and Its Hidden Mystical Language.* Kessinger Publishing's Rare Reprints.

Lao Tzu. *Tao Te Ching.* Trans. Victor H. Mair. NY: Bantam, 1990.

Ling Shu, or The Spiritual Pivot. Trans. Wu Jing-Nuan. Washington DC: Taoist Center, 1993 (paperback ed. 2002).

Luomala, Katharine. *Maui of a Thousand Tricks: His Oceanic and European Biographers.* Bernice P. Bishop Museum Bulletin 198. Honolulu: Bernice P. Bishop Mueseum, 1949.

Mahabharata. Trans. Kisari Mohan Ganguli. 1883 to 1896. www.sacred-texts.com.

Mathisen, David Warner. *The Mathisen Corollary: Connecting a Global Flood with the Mystery of Mankind's Ancient Past.* Paso Robles, California: Beowulf Books, 2011.

Mathisen, David Warner. *Star Myths of the World, and how to interpret them, Volume One.* Paso Robles, California: Beowulf Books, 2015.

Mathisen, David Warner. *Star Myths of the World, and how to interpret them, Volume Two.* Paso Robles, California: Beowulf Books, 2016.

Mathisen, David Warner. *Star Myths of the World, and how to interpret them, Volume Three.* Paso Robles, California: Beowulf Books, 2016.

Mathisen, David Warner. *The Undying Stars: the truth that unites the world's ancient wisdom, and the conspiracy to keep it from you.* Paso Robles, California: Beowulf Books, 2014.

Neihardt, John G. *Black Elk Speaks: Being the Life Story of a Holy Man of the Oglala Sioux*, Premiere Ed. Albany: SUNY Press, 2008.

Neugebauer, Otto. *The Exact Sciences in Antiquity.* Second ed. NY: Dover, 1969.

Neugebauer and Parker. *Egyptian Astronomical Texts.* 3 vols. Providence, Rhode Island: Brown UP, 1968.

Patañjāli. *Yoga Sutras.* Trans. Ronald Steiner. http://www.ashtangayoga.info/source-texts/yoga-sutra-patanjali/chapter-1/

Pausanias. *Description of Greece.* Trans. J. G. Frazer. 2nd ed. London: MacMillan, 1913.

Pindar. *Odes.* Trans. Diane Arnson Svarlien. Perseus Digital Library: Tufts University, 1990. perseus.tufts.edu

Plutarch. *De esu carnium*, in Moralia. Online edition of the Loeb Classical Library, 1957. www.penelope.uchicago.edu.

Plutarch. *De Iside et Osiride*, in *Moralia*. Online edition of the Loeb Classical Library, 1957. www.penelope.uchicago.edu.

Rey, H. A. *The Stars: A New Way to See Them.* Boston: Houghton-Mifflin, 1952. Enlarged World-Wide Edition, 1988.

Rolleston, Frances. *Mazzaroth; Or, the Constellations.* London: Rivingtons, 1862.

Schoch, Robert. *Forgotten Civilization: The Role of Solar Outbursts in our Past and in our Future.* Rochester, Vermont: Inner Traditions, 2012.

Schwaller de Lubicz, R. A. *Esoterism & Symbol.* Originally published in French under the title *Propos sur Esotérisme et Symbole* by La Colombe, Editions du Vieux Colombier, in 1960. English translation by André and Goldian VandenBroeck, 1985. Rochester, Vermont: Inner Traditions, 1985.

Seiss, Joseph A. *The Gospel in the Stars; or, Primeval Astronomy.* Philadelphia: Claxton, 1882.

Sellers, Jane B. *Death of Gods in Ancient Egypt: A Study of the Threshold of Myth and the Frame of Time.* Lexington, Kentucky: Lulu, 1992. Rev. Ed. 2007.

Shakespeare, William. *Hamlet.* in *The Norton Shakespeare: Based on the Oxford Edition.* Stephen Greenblatt, Gen. Ed. NY: Norton, 1997. 1659 - 1759.

Shakespeare, William. *Macbeth.* in *The Norton Shakespeare: Based on the Oxford Edition.* Stephen Greenblatt, Gen. Ed. NY: Norton, 1997. 2555 - 2618.

Smith, George. *The Chaldean Account of Genesis.* NY: Scribner, Armstrong & Co., 1876.

Sturluson, Snorri. *Edda.* Trans. Anthony Faulkes. London: Everyman, 1987.

Taylor, Robert. *Astronomico-Theological Lectures.* NY: Calvin Blanchard, 1857.

Taylor, Robert. *Devil's Pulpit: or, Astro-Theological Sermons by the Reverend Robert Taylor.* NY: Calvin Blanchard, 1857.

Temple, Robert K. G. *Sirius Mystery.* Rochester, Vermont: Destiny, 1976. Reprint 1987.

Trant, Captain Thomas Abercromby. *Narrative of a Journey Through Greece, in 1830.* London: Henry Colburn and Richard Bentley, 1830.

Wallis Budge, Ernest A. *Osiris and the Egyptian Resurrection, Vol 2.* NY: Putnam, 1911.

Waters, Frank. *Book of the Hopi.* NY: Viking Penguin, 1963. Paperback Penguin Books paperback ed., 1977.

West, John Anthony. *Serpent in the Sky: The High Wisdom of Ancient Egypt.* 1987 ed. NY: Julian Press, 1987.

Van Gulik, R. H. *Sexual Life in Ancient China: A preliminary survey of Chinese sex and society from ca. 1500 BC till 1644 AD.* Leiden: EJ Brill, 1961. Reprint 1974.

Volney, Constantin François de Chasseboeuf. *Volney's Ruins: or, Meditation on the Revolutions of Empires.* Translated under the immediate inspection of the author from the latest Paris edition. Boston: Josiah P. Mendum, 1869.
https://archive.org/stream/volneysruinsormoovoln#page/n5/mode/2up

Zeitgeist (movie), part 1. Peter Joseph. 2007.

Index

Aborigines, or Aboriginal Australians 24
Abraham 142
Abydos 77
Achaeans 71, 81, 111, 225 - 226, 275
Achilles 32, 106 - 107, 113, 115 - 116, 121, 192, 225
Actaeon (or Acteon) 32 - 33
Adam and Eve 11 - 12, 28, 126
Aesir 142 - 143
Africa 61, 67, 256
Akawaio 134, 136
Akkad 27, 127, 132 - 133
Albertus Magnus (c. 1200 - 1280) 12
Alcmene (or Alcmena) 192
Aldebaran 97
Alibamu 164
Alison, Jim 245
Amaterasu 146 - 147, 149, 151, 153
Ame-no-Uzume-no-mikoto 148, 151, 153
Amphion 192, 203
Amphitrite 219
Amphitryon 192
Amun 254
Anacalypsis (1836) 233
Andromeda 277 - 281
Annual cycle 5, 49 - 59, 64, 71, 74, 80, 86 - 88, 92, 97, 156, 196 - 197, 238
Anointing 118 - 119, 131
Aotearoa 105, 108
Aphrodite 214, 275
Apocalyptic traditions 68, 70, 123
Apollo 107, 195 -196, 214, 262, 271 - 276, 278 - 279
Apollodorus (see Pseudo-Apollodorus)
Apples 142, 276

Apuleius (c. AD 124 - c. AD 170) 213 - 214, 216, 275
Aqqi 128
Aquarius 102, 129, 200, 258 - 259
Aquila 119, 149, 173 - 174
Arcturus 152
Aries 120, 188 - 189
Argo Navis 134, 137
Aristophanes (444 BC - 385 BC) 155
Ark 28, 127, 131 - 135
Artemis 32 - 33, 118, 278
Asgard 143
Ashvins 192
Astartĕ 117
Astromythology 9
Astronomico-Theological Lectures 8 - 9, 11
Astrotheology 5 - 7, 9, 98, 235, 239, 250
Athena 194, 219, 221, 275
Athenaïs 117
Athens 155
Atman 208 - 209, 211
Atwill, Joseph 255
Atrahasis 133
Aum (see Om)
Australia 16, 61, 67, 256
Axis 45 - 47, 49, 63, 81, 85, 87 - 88, 90 - 93, 237 - 238
Baalbek 245
Babylon 10, 78
Balaam 279
Baldr (or Balder) 143, 180 - 182
Bandicoot Woman 24
Baptism 4, 105 - 108, 115 - 117, 120 - 121, 124 - 125, 139, 145, 157
Barbiero, Flavio 255
Baskets 27, 128, 130 - 132, 134 - 137, 140, 252
Baubo 145

Bede (c. 672 - 735) 201
Bethel 234
Beyond Vietnam 289 - 291
Bhagavad Gita 152, 210, 269, 292
Bhāve, Vinobā (1895 - 1982) 230
Bhima 41 - 42
Bhoktr 210
Bible 3, 5, 7 - 13, 15, 17, 67, 102,
   106, 113, 120, 124, 131,
   133 - 134, 151, 177, 180, 241,
   254, 257, 264, 292
Black Elk (Heháka Sápa)
   (1863 - 1950) 185- 186,
   263 - 265, 270, 284,
   287 - 288, 291
Blackfoot 164
Blessing versus cursing 257 - 258,
   267, 282 - 283
Bonacci, Santos 235
Boötes 151 - 153, 155, 175
Bora Bora 111
Bow-trial 99
Brahman 293
Buddhism 73
Buckley, Theodore Alois
   (1825 - 1856) 100, 219
Burley, Paul 251 - 252
Byblos 77, 98, 116, 118, 145
Caesar 196
Calliope 159
Calypso 217, 239
Cancer 62 - 63, 76
Cantonese 73, 227, 236
Capricorn 76, 199 - 200, 242,
   258 - 259
Castor and Polydeuces
   (or Castor and Pollux)
   192 - 194, 200 - 201, 204,
   217
Carlson, Randall 255
Celeus 144, 146
Cerberus 165, 174
Chakras 64, 122, 225, 235

Chamberlain, Basil Hall
   (1850 - 1935) 161
Chanting 227 - 229, 230, 233, 242
Cherokee 164
Chi (or *Hei*) see *Qi*
China 71, 73, 123, 228, 236, 241,
   244, 249
Chinook 164
Christianity 6, 12 - 13, 106 - 107,
   155, 180, 254, 256 - 257,
   269 - 270, 288 - 289
Circe 239
Columba 134
Colonialism 256, 270, 290
Coma Berenices 151
Comanche 169, 176, 179
Corê (see Persephone)
Corona Australis
   (Southern Crown) 15, 103,
   105, 164
Corona Borealis
   (Northern Crown) 36, 38,
   101 - 107, 128, 139, 161
Corvus 271
Crazy Horse (Tashunke Witkó)
   (c. 1842 - 1877) 185 - 186
Cruttenden, Walter 93, 297 n.26
Cupid (see Eros)
Curious George 19
Cursing (see "blessing versus
   cursing")
Cutimbo 251
Cuzco 251
Cyclops 220 - 221, 239
Cygnus 119, 149, 173 - 174, 193
Cypselus (d. 627 BC) 201
Daikoku 153
Daily cycle 45 - 46, 50 - 51, 53 - 54,
   63, 68, 85 - 87, 92, 97
Dakota Access Pipeline 264
Damascus 195 - 196
David 171, 192, 195
Daytime 46 - 47, 63, 90

de Brébeuf, Jean (1593 - 1649) 167, 180
de Santillana, Giorgio (1902 - 1974) 14, 16 - 18, 96, 244
Decans 82
Delphi 25, 230, 232 - 233, 254
Delphinus 173
Demeter 143 - 146, 155 - 157, 272, 284
Demetrius of Phalerum (350 BC - 280 BC) 231
Demophon 146
Deucalion 139
Devil's Pulpit 8, 9, 11, 195
Dhritarastra 224
Diagonal Calendars 82 - 83
Didymus (see Thomas)
Diodorus Siculus (first century BC) 144
Dionysus (or Dionysos) 29, 32, 272, 274, 276, 282, 284
Dioscuri 201, 203
Djed 4, 75 - 77, 81 - 104, 120, 122, 140, 197, 199
Dodona 231
Draupadi 119
Drums 256
Dumuzi (or Tammuz) 182
Dunn, Christopher 244
Dupuis, Charles François (1742 - 1809) 10 - 11
Durga 293 - 295
Easter: The Birthday of the Gods 55
Easter Island (see Rapa Nui)
Ecliptic 54, 60
Edda 143, 180
Elements 38, 51 - 52, 54, 156
Eleusinian Mysteries 144 - 145, 155 - 156, 254

Egypt 4 - 5, 10, 53, 56, 69 - 70, 74 - 76, 81 - 82, 84 - 85, 92, 96, 121, 127, 182, 198, 231, 244 - 247, 249, 254, 255
Equinoxes 5 - 6, 14, 54 - 60, 65, 74 - 76, 86 - 87, 92 - 93, 95 - 97, 122, 156, 188 - 189, 202
Enkidu 40 - 41, 78, 137, 140, 192, 203
Enlil 138 - 139
Enoch 12
Entheogens 256
Eros 206, 213 - 217, 221, 287
Esau 192, 203
Esoteric 2 - 3, 6 - 7, 13, 57, 66 - 67, 76, 155 - 156, 203, 213, 216, 244, 275
Eurydice 159 - 160, 163, 165, 168, 170 - 171, 178, 190
Eve 11 - 12, 28, 126
Exodus 10, 127, 233
Ezekiel 12, 182
Faiia, Santha 251
Farrell, Joseph P. 244, 248
Flavors 238
Fox (Native American nation) 164
Fox (animal) 250, 280
France 10 - 12
Freke, Timothy 255
Freud, Sigmund (1856 - 1939) 211
Freya or Freyja 152
Fukushima 285
Galactic Core 62 - 63, 172, 188, 44, 250
Gandy, Peter 255
Gawain 218
Gayton, Anna H. (1899 - 1977) 163, 165 - 166, 168, 172, 182
Gemini 62 - 63, 76, 96 - 97, 194 - 196, 198, 217
Genesis 52, 128, 132 - 133, 138, 142, 149, 157, 179, 235, 241 - 242, 284

Genetically-modified food 284 - 288
Gephrysmoi 155
Gilgamesh 15, 28, 40 - 41, 78, 133, 137 - 140, 192, 250
Gioll (see Gjöl)
Giovanni del Giglio (d. 1554) 199
Giza 56, 245, 247
Gjallarbrú 180
Gjöl (or Gioll) 180
Gnosis 182, 270
Gnostics 255
Göbekli Tepe 246 - 248, 250 - 251
Golden Ass (Apuleius) 213 - 214, 275
Gospel in the Stars (1884) 12 - 13
Graves, Robert (1895 - 1985) 231, 233
Great Rift 63, 72, 172, 174, 188, 250 - 251
Great Sphinx (see Sphinx at Giza)
Great Square 134, 137, 258 - 259
Greece 17, 25, 64, 67, 69 - 71, 78, 81, 84, 100, 115, 117 - 118, 121, 139, 143, 146, 152, 157, 159 161, 163, 165, 168 - 170, 172, 174 - 175, 181 186, 192 - 193, 195 - 196 224, 226, 253 - 254, 266, 269, 272, 274, 278 - 280
Green Knight 218
Hades 161 - 164, 177
Hahaki 161
Hall, Sidney (1788 - 1831) 21
Ham 257 - 260, 266
Hamlet 121, 139
Hamlet's Mill (1969) 14 - 17, 69, 96, 98, 134 - 135, 137, 253 - 254, 287
Hancock, Graham 244 - 245, 248 - 251, 255, 260, 264

Hanuman 41 - 43, 135
Harris, James Rendel 200 - 203
Hatshepsut (reign 1478 - 1458 BC) 83
Hawai'i 105, 111
Hawaiki 111
Heart 66, 121, 177, 208, 237 - 238, 174
Hector 225 - 226
Heháka Sápa (see Black Elk)
Heimskringla saga 201
Hel 180 - 181
Helen 275
Heliacal rise 82, 91, 96
Hephaestus 279
Hera 274
Heracles 38, 40, 192 - 193
Hercules 35 - 43, 99, 109, 110, 112 - 115, 122, 124, 136 - 137, 152, 171, 175 - 177, 179, 192 - 193, 234
Hermetic tradition 187
Hermion 144
Hermod 180 - 181
Herodotus (c. 5th century BC) 155, 231
Higgins, Godfrey (1772 - 1833) 233
Higher Self 65, 190 -191, 194, 196 - 197, 199 - 200, 203 - 204, 207, 211, 213, 216 - 217, 223, 225 - 226, 242, 268 - 269, 282, 294
Hiruko 27, 126 - 130
Horus 5, 53, 82
Huang Di Nei Jing 236
Hultkrantz, Åke 168 - 169, 172, 183
Hunahpu 192
Hunkpapa 263
Huron 164, 167
Hyades 15, 279
Iambe 144 - 146, 155

Idzumo 161
Iliad 68, 73, 79, 81, 100, 107, 152, 175, 192, 210, 224 - 226, 233, 271
Ilium 71, 226, 233, 235, 274
Imperialism 256 - 257, 290
India 27, 41, 64 - 65, 71 -72, 81, 119, 127, 135, 152, 168, 207, 224, 228, 235, 256, 266
Indus-Saraswati Civilization 244, 249, 255
Ino 218
Interior Planets 47
Iowa 233
Iphicles 192 - 193
Ishvara 228
Isis 76 - 77, 82 - 84, 97, 116, 117 - 119, 145, 214, 275
Iyengar, B. K. S. (1918 - 2014) 209, 230
Izanagi and Izanami 125 - 126, 128 - 130, 161, 163, 171, 178
Jacob 192, 203, 234 - 235
Jacob's Ladder 234 - 235
Japan 27, 70, 125 - 127, 146, 148, 153, 161, 163, 165, 168 - 169, 171 - 172
Japheth 257 - 260, 266
Jawbone 15, 279
Jefferson, Thomas (1743 - 1826) 11
Jesus 5, 13, 56, 77, 179, 192, 195, 198 - 200, 203, 213, 220, 240, 256, 266 - 267, 270, 287
John the Baptist 102
Jonathan 192, 195
Jove 232
Judgment of Paris 273 - 276
Judgment of Solomon (see Solomon)
Julius Caesar (see Caesar)
Kabiri (or Kabeiroi) 203
Kami 125, 151, 161
Karna 27, 127, 135
Kastor (see Castor and Polydeuces)
Katha Upanishad (or Kathopanishad) 207, 209, 211, 219, 269
Kauravas 71, 81, 224, 285
Kehinde 192
Kephissos 155
Kidneys 169, 176, 237 - 238
Kildeer 165, 174
King, Martin Luther, Jr. (1929 - 1968) 289 - 292
King James translation 196
Kingsley, Peter 267 - 269, 283, 287
Koasati 164
Kojiki 27, 70, 125, 127 - 128, 130, 146 - 147, 149, 151, 161, 165, 171 - 172, 178
Krishna 152, 210 - 211, 269, 287, 292 - 294
Kuhn, Alvin Boyd (1880 - 1963) 7, 18, 51 - 53, 55, 57 - 58, 63, 71, 74 - 76, 79, 87, 103, 121, 197, 202 - 204, 234, 293
Kundalini 64, 140, 226
Kunti 127, 224
Kurukshetra 71, 210, 243, 292, 294
Kwakiutl 163
Lakota 185, 263, 270
Leap year 95
Leda 193
Lehner, Mark 246
Leo 24, 88
Leopards 24
Lepus 97
Leucotheia (or Leucothea) 218
Libra 26, 35, 56 - 57, 231
Light on Yoga 209, 230
Ling Shu (see *Huang Di Nei Jing*)
Lions 24, 40 - 41, 69, 84, 139, 288, 292

317

Liver 237-238
Loki 143, 151-153, 181
Lot's wife 179-180, 189
Lotus-Eaters 239
Lungs 237-238
Luomala, Katharine (1907-1992) 105, 108, 111
Lycophron 107
Lyra 171
Macrocosm 233, 236-237, 242
Madri 224
Magi 11
Mahabharata 27, 41, 67-68, 73, 79, 81, 119, 127, 152, 177, 210, 224, 243, 269, 285, 292
Makunaima 134
Malcander 117
Maltwood, Katharine Emma (1878-1961) 244
Mair, Victor H. 71
Maithuna 226, 242
Malecite 164
Mantagnais 164
Mantras 228
Mars 237
Marsyas 272, 278
Martial arts 123, 222, 242
Mary 77, 179
Maui 26-27, 71, 105-108, 110-113, 115-116, 122, 125-126
Maya 24, 56, 64, 72, 161, 188, 192, 251, 256, 269
Mazzaroth (1862) 12
Meditation 56, 122, 124, 140, 222, 226-228, 242, 269, 287
Menomini 164
Mercury 47-48, 224, 237
Metamorphoses of Apuleius (see "Golden Ass")
Metanira 146
Michell, John (1933-2009) 244
Micmac 164

Microcosm 225, 233, 235-238, 242
Midas 272-279, 282-283, 285-286, 288, 291-292
Milky Way 15, 27-28, 35, 62-64, 69, 72, 110-113, 118-119, 122, 128-129, 131, 137, 140, 149, 151, 171-174, 182, 188, 193, 234, 242, 250, 277-278
Mist-painting 17, 253-254, 287
Miwok 164
Moai 251
Moðguðr 180
Modoc 164
Mono 164
Moon 2, 5-7, 9-10, 14, 46, 49, 65-66, 85, 233, 293
Moorea 111
Morris, William (1834-1896) 297 n. 14
Moses 27, 127-128, 130, 132, 134-135, 140-141, 173
Mount Hiba 161
Mount Timolus (see Timolus)
Muses 159
"Myriad Things" (萬物) 71-72, 79
Mysteria 155-156, 203, 214, 254
Nachiketa 207-208, 211
Nag Hammadi 198
Nausicaa 118
Navajo 166
Nei Daan 226
Nei Gong 226
Neihardt, John G. (1881-1973) 263
Nemanûs 117
Neptune (see Poseidon) 100, 219
Neugebauer, Otto 84
New Zealand (see Aotearoa)
Njord (or Niord) 143
Noah 128, 132, 134, 138-139, 257-259

Northern Crown (see Corona Borealis)
Nutka 163
Ocean Stream 27
Odin 180
Odyssey 64, 68, 73, 78 - 79, 98 - 103, 119, 139, 179, 217 - 221, 239 - 240, 271
Odysseus 34 - 35, 78 - 79, 98 - 103, 119, 139, 179, 217 - 221
Oedipus 27
Oglala 185, 263
Ogygia 98
Oicotype 4, 172, 177 - 180, 183, 188, 273
Om 228 - 230, 233
Omphalos 232 - 233
Onan 241 - 241, 301 n.125
Ophiucus 28, 119, 130 - 131, 134 - 137, 149, 170 - 172, 174 - 177, 179, 182, 234, 250
Oracle at Delphi (see Delphi)
Origin of All Religious Worship (1795) 10 - 11
Organs 237 - 238
Orpheus 159 - 161, 163, 165, 168 - 172, 178, 188, 190
Oscotarach ("Pierce-Head") 167, 175 - 176
Osiris 53, 74 - 77, 79, 81 - 82, 84 - 88, 92, 96 - 99, 101 - 103, 116, 119 - 120, 122, 182
Otafuku 153
Pacific Islands 26, 67, 105, 108, 111, 115, 121, 251, 255 - 256, 263
Pactolus 276 - 277, 283, 291
Pan 272, 278
Pan-pipes 272, 278, 280 - 281
Pan Painter 32
Pandavas 41, 71, 81, 224, 294
Pandu 224
Paris 107, 273 - 276
Patañjāli 122, 228
Paul 194 - 199

Pausanius (c. AD 110 - AD 180) 201
Pawnee 164
Pegasus 134, 137, 258
Peleus 107
Penelope 79, 103, 239
Pentateuch 10
Persephone 143 - 146, 156 - 157, 159, 163, 177, 216
Peru 251
Phaeacia 220
Pindar (c. 517 BC - c. 437 BC) 193, 229
Pisces 189, 258
Pitch 12 - 128, 131 - 132
Plato (c. 420s BC - 340s BC) 17, 254
Planets 2, 14, 46 - 19, 66, 85, 87 - 90, 93 - 94, 224, 233, 237 - 238, 245, 249, 253, 258, 260, 271, 285
Plotinus (c. AD 205 - 270) 53
Plutarch (AD 45 - AD 120) 76, 97, 116 - 117, 119, 155, 230, 232, 254, 284, 298, 302
Pollux (see Castor and Polydeuces)
Polydeuces (see Castor and Polydeuces)
Polyphemus 220, 239
Pomegranate 163, 177
Poseidon 100, 139, 217, 219, 221, 272
Prana 64, 227, 229
Praṇava 230
Praxithea 146
Precession 14, 65, 68, 92 - 98, 120, 228, 245, 300
Precessional constant 95, 97 - 98
Precessional numbers 95, 97 - 98, 228, 245
Priam 107
Prodigal Son 203 - 204
Prometheus 181

319

Proverbs 211, 300 n.98
Pseudo-Apollodorus 107, 144 - 145
Psyche 206, 213 - 219, 221, 269, 287
Puma Punku 245
Pythagoras / Pythagoreans 17
Pythia 25
Qi (or *Hei*) 227, 229, 238 - 240
Qigong 140
Quantum Physics 184
Ragnarök 78
Raiatea 111
Ramayana 67, 73, 269
Rapa Nui (Easter Island) 251
Red Sea 173, 233 - 234
Remus 192
Revelation 15, 27, 68 - 70, 123 - 124
Rey, H. A. (1898 - 1977) 19 - 26, 28 - 29, 31 - 32, 34 - 38, 40, 42 - 43, 110, 112, 115, 150, 251 - 252
Rey, Margret (1906 - 1996) 18 - 19, 43
Rolleston, Frances (1781 - 1864) 12 - 13
Rome 168, 192, 195, 236
Roman Empire 236, 254 - 255, 265, 269, 288 - 290
Romulus 192
Rubens, Peter Paul (1577 - 1640) 114
Ruins: or, Meditations on the Revolutions of Empires (1791) 11
Russia 289
Sabaism 11
Sacrum 235
Sagittarius 15, 29 - 30, 32 - 35, 62 - 63, 69, 72, 76, 110, 113, 118 - 121, 128 - 129, 131, 170 - 172, 176, 178, 180, 188 - 189, 193, 195 - 197, 199, 225, 234, 242, 250 - 252, 259, 277 - 279

Sah or Sahu 84
Samothrace 203 - 300
Samson 15, 151, 279, 297
Sanskrit 27, 41, 65, 67, 72 - 73, 127, 177, 210, 227, 230
Saosis 117
Sarah 142, 149, 156 - 157
Sargon 27, 127 - 128, 132, 135
Saturn 14, 237
Satyrs 270, 272 - 273, 278 - 282
Saul 195 - 199
Saxo Grammaticus (c. 1150 - 1220) 201
Scapulamancy 126
Schmidt, Klaus (1953 - 2014) 246 Schoch, Robert 244 - 246, 248 - 251, 255, 260, 301
Schwaller de Lubicz, R. A. (1887 - 1961) 66, 244, 297
Scorpio (or Scorpius) 15, 26 - 28, 33, 35, 62 - 63, 69, 71 - 72, 110 - 112, 118 - 123, 128 - 130, 134, 137, 170 - 172, 174, 176, 193, 234, 250 - 277
Seasons 49, 54 - 55, 74 - 76, 81, 86 - 87, 196, 236 - 237
Second birth 64, 121, 202 - 203
Seed Realm 184 - 185, 229, 291
Seiss, Joseph A. (1823 - 1904)
Sellers, Jane B. 96 - 97
Seneca 164
Senenmut 83
Set or Seth 97, 116, 254
Seth, son of Adam 12
Seti I (reign 1290 BC - 1279 BC) 83
Shakespeare 139
Shamanism 140, 183, 256, 266 - 267, 270
Shasta 164
Shem 257 - 260, 266
Sheol 196
Sif 151

320

Sigu 134 - 137, 139, 252
Silenus 281
Sirius 82
Sirius Mystery 230 - 231, 233
Sitting Bull (Tatánke Íyotanke) (c. 1831 - 1890) 263
Skade (or Skadi) 142 - 143, 151, 153, 156 - 157
Skalskaparmal 143
Smartphone apps 29
Snorri Sturluson (1179 - 1241) 143, 180 - 181
Sodom 179, 189
Solstices 5 - 6, 54 - 55, 57 - 58, 60, 63, 66, 74 - 76, 87, 95, 120 - 122, 188, 199, 202, 251
Solomon 113 - 116, 177, 273 - 275, 282
Society Islands 111
Southern Crown (see Corona Australis)
Sphinx at Giza 245 - 247
Speculum Astronomiae 12 - 13
Standing Rock 263
Stars, The: A New Way to See Them (1952) 19 - 43, 150
Statius 107
Strabo (65 BC - AD 23) 155
Styx 106 - 107, 113, 165
Sumer 10, 78, 133, 270
Sun 2, 5 - 10, 14, 45 - 63, 66, 81 - 82, 85 - 92, 94 - 96, 120, 188 - 189, 195 - 197, 215, 224, 233, 237 - 239, 251 - 252, 276, 293
Susanowo (or Susano'o) 146
Svarlien, Diane Arnson 193
Taboo (or tabu) 178
Taiowa 232
Taiwo 192
Talmud 201
Tama 108, 113, 122
Tammuz (see Dumuzi)
Tantra 123 - 124, 226, 242

Tao Te Ching (or Dao De Jing) 71 - 73, 79, 221, 266, 269
Tat (see Djed)
Tatánke Íyotanke (see Sitting Bull)
Taurus 15, 97, 297
Tawakoni 164
Taylor, Robert (1784 - 1844) 7 - 11, 18, 195 - 197, 297
"Teapot" 31, 250 - 251
Telumni 165
Temple, Robert 230 - 233
Thebes 192, 231
Theodosius (AD 347 - AD 395) 155, 254
Thesmophoria 144
Thetis 107, 113
Thomas 192, 196 - 200, 212 - 213, 216, 219 221, 229, 269, 287
Thor 152, 269
Thrace 159
Timolus 273, 276
Tiresias 99 - 101
Tissot, James Jacques Joseph (1836 - 1902) 115
Tjasse (or Thiassi) 142, 156
Tlingit 163
Trojans 71, 81, 225 - 226
Trojan War 71, 81, 225 - 226, 273, 275
Turning-point 55 - 57, 63 - 64, 188 - 189, 202
Twins and twinning 191 - 192, 194 - 195, 197 - 205, 207, 210 - 211, 213, 217, 223, 226 - 227, 265
Two Bears, Cody 264
Tyndareus 193
Typhon 254
Underworld 52 - 54, 59, 75, 81, 56 - 84, 92, 103, 105, 143, 159 - 161, 163 - 167, 169 - 176, 178, 181 - 182, 188 - 190, 216
United States 232, 290

Undying Stars (2014) 15, 120, 184, 196, 255
Urania's Mirror (1825) 21, 28, 35
Uta-napishti (or Utnapishtim) 133, 137 - 140
Uzume 148, 151, 153
van Gulik, Robert 241 - 242
Vayu 227
Vedas 192, 250, 266
Venus 47 - 48, 214, 216, 224, 237
Vietnam 289 - 291
Vindemiatrix 24, 154
Virgo 21 - 29, 35, 110 - 111, 123, 128 - 129, 131, 148, 150 - 154, 170 - 171, 177, 188 - 189, 271
Volney (Constantin François de Chasseboeuf, Comte de Volney) (1757 - 1820) 10 - 11
von Dechend, Hertha (1915 - 2001) 14, 16, 18, 96, 244
von Sydow, Carl Wilhelm (1878 - 1952) 4
Wallis Budge, Ernest A. (1857 - 1934) 84
Wasi'chu 263 - 264, 267, 270, 287, 291
Waters, Frank (1902 - 1995) 232
Watkins, Alfred (1855 - 1935) 244
West, John Anthony 244 - 245, 248
Whirlwind 176 - 177
White Bear (Oswald Fredericks) (1905 - 1996) 232
Wichita 164
Wikipedia 29
Winnebago 164
Winnowing fan 100 - 102
World War II 43
Xbalanque 192
Yama 207 - 209, 211
Yeh Tê-hui (1864 - 1927) 241 - 242
Yggdrasil 78
Yoga 121 - 124, 140, 209, 222, 228 - 230, 242, 287

Yoga Sutras 121, 228 - 229
Yokuts 164 - 165
Yomo (or Yomi) 161
Yomo-tsu 162
Yoruba 192
Yuchi 164
Zeitgeist (2007) 5
Zethus 192
Ziusudra 133
Zodiac 12, 26, 29, 35, 55, 58 - 60, 62, 64, 72, 76, 87, 148, 188 - 189, 194, 196, 199, 202, 223, 225, 238, 242
Zuñi 164

CPSIA information can be obtained
at www.ICGtesting.com
Printed in the USA
BVOW03s1117110617
486274BV00004B/110/P